Differential Equations
with *Maple*™
Second Edition

Revised for Maple V Release 4

Kevin R. Coombes
Brian R. Hunt
Ronald L. Lipsman
John E. Osborn
Garrett J. Stuck

All of the University of Maryland at College Park

John Wiley & Sons, Inc.
New York • Chichester • Weinheim • Brisbane • Singapore • Toronto

ISBN 0-471-17645-1

Printed in the United States of America

10 9 8 7 6 5 4 3 2

Printed and bound by Malloy Lithographing, Inc.

Preface

As the subject matter of differential equations continues to grow, as new technologies become commonplace, as old areas of application are expanded, and as new ones appear on the horizon, the content and viewpoint of courses and their textbooks must also evolve.

Boyce & DiPrima, **Elementary Differential Equations**, Sixth Edition.

Traditional introductory courses in ordinary differential equations (ODE) have concentrated on teaching a repertoire of techniques for finding formula solutions of various classes of differential equations. Typically, the result was rote application of formula techniques without a serious qualitative understanding of such fundamental aspects of the subject as stability, asymptotics, dependence on parameters, and numerical methods. These fundamental ideas are difficult to teach because they have a great deal of geometrical content and, especially in the case of numerical methods, involve a great deal of computation. Modern mathematical software systems, which are particularly effective for geometrical and numerical analysis, can help to overcome these difficulties. This book changes the emphasis in the traditional ODE course by using a mathematical software system to introduce numerical methods, geometric interpretation, symbolic computation, and qualitative analysis into the course in a basic way.

The mathematical software system we use is *Maple*. (This book is also available in a *Mathematica* version.) We assume that the user has no prior experience with *Maple*. We include concise instructions for using *Maple* on three popular computer platforms: *Windows, Macintosh,* and the *X Window System*. This book is not a comprehensive introduction or reference manual to either *Maple* or any of the computer platforms. Instead, it focuses on the specific features of *Maple* that are useful for analyzing differential equations. It also describes the features of the *Maple* "Worksheet" interface that are necessary for creating a finished document.

This supplement can easily be used in conjunction with most ODE texts. It addresses the standard topics in ODE, but with a substantially different emphasis.

We had two basic goals in mind when we introduced this supplement into our course. First, we wanted to deepen students' understanding of differential equations by giving them a new tool, a mathematical software system, for analyzing differential equations. Second, we wanted to bring students to a level of expertise in the mathematical software system that would allow them to use it in other mathematics, engineering, or science courses. We believe that we have achieved these goals in our own classes. We hope this supplement will be useful to students and instructors on other campuses in achieving the same goals.

Acknowledgements

We thank Peter Olver, Jonathan Rosenberg, Larry Shampine, Shagi-Di Shih, and Nancy Stanton, all of whom contributed to this book. We are grateful to the 35 or so of our colleagues who have used the first edition and other versions of this material in their classes at the University of Maryland at College Park. We are particularly indebted to the many students who have used these materials, and have communicated to us their comments and suggestions. Finally, we thank Barbara Holland, mathematics editor at Wiley, for her enthusiastic and thoughtful support of our project.

Kevin Coombes
Brian Hunt
Ron Lipsman
John Osborn
Garrett Stuck

College Park, Maryland
September 30, 1996

Contents

Chapter 1

Introduction

We begin by describing the philosophy behind our approach to the study of ordinary differential equations. This philosophy has its roots in the way we understand, use, and apply differential equations; it has influenced our teaching and guided the development of this book. This chapter also contains two user's guides, one for students and one for instructors.

Guiding Philosophy

In scientific inquiry, when we are interested in understanding, describing, or predicting some complex phenomenon, we use the technique of *mathematical modeling*. In this approach, we describe the state of a physical, biological, economic, or other system by one or more functions of one or several variables. For example, the position $s = s(t)$ of a particle is a function of the time t; the temperature $T = T(x, y, z, t)$ in a body is a function of the position (x, y, z) and the time t; the gravitational or electromagnetic force on an object is a function of its position; the money supply is a function of time; the populations $x = x(t)$ and $y = y(t)$ of two competing species are functions of time. Next, we attempt to formulate, in mathematical terms, a fundamental law governing the phenomenon. Typically, this formulation results in one or more differential equations; *i.e.*, equations involving derivatives of the functions describing the state of the system with respect to the variables they depend on. Frequently, the functions depend on only one variable, and the differential equation is called *ordinary*. To be specific, if $x = x(t)$ denotes the function describing the state of our system, an *ordinary differential equation* in $x(t)$ might involve $x'(t)$, $x''(t)$, higher derivatives, or other known functions of t. By contrast, in a *partial differential equation*, the functions depend on several variables.

In addition to the fundamental law, we usually describe the initial state of the system. We express this state mathematically by specifying *initial values* $x(0)$, $x'(0)$, *etc*. In this way, we arrive at an *initial value problem*—an ordinary differential equation together with initial conditions. If we can solve the initial value problem, then the solution is a function $x(t)$ that predicts the future state of the system. Using the solution, we can describe qualitative or quantitative properties of the system. At this stage, we compare the values predicted by $x(t)$ against experimental data that we accumulate by observing the system. If the experimental data and the function values match, we congratulate ourselves on a job well done and go on to the next challenge. If they do not match, we go back, refine the model, and start again. Even if we are satisfied with the model, we may later arrive at a time when new technology, new requirements, or newly discovered features of the system render our old data obsolete. Again, we respond by reexamining the model.

The subject of differential equations consists in large part of building and solving mathematical models. Many results and methods have been developed for this purpose. These results and methods fall within one or more of the following themes:

(1) Existence and uniqueness of solutions

(2) Dependence of solutions on initial values

(3) Derivation of formulas for solutions

(4) Numerical calculation of solutions

(5) Graphical analysis of solutions

(6) Qualitative analysis of differential equations and their solutions.

The basic results on existence and uniqueness (theme 1), and dependence on initial values (theme 2), form the foundation of the subject of differential equations. The derivation of formula solutions (theme 3) is a rich and important part of the subject; a variety of methods have been developed for finding formula solutions to special classes of equations. Although many equations can be solved exactly, many others cannot. However, any equation, solvable or not, can be analyzed using numerical methods (theme 4), graphical methods (theme 5), and qualitative methods (theme 6). The ability to obtain qualitative and quantitative information without the aid of an explicit formula solution is crucial. That information may suffice to analyze and describe the original phenomenon (which led to the model, which gave rise to the differential equation).

Traditionally, introductory courses in differential equations focused on methods for deriving exact solutions to special types of equations, and included some simple numerical and qualitative methods. The human limitations involved in compiling numerical or graphical data were formidable obstacles to implementing more advanced qualitative or quantitative methods. Computer platforms have reduced these obstacles. Sophisticated software and mainframe computers have enhanced the use of quantitative and qualitative methods in the theory and applications of differential equations. With the arrival of comprehensive mathematical software systems on personal computers, this modern approach has become accessible to undergraduates.

In this book, we use the mathematical software system *Maple* to implement this new approach. We introduce symbolic, numerical, graphical, and qualitative techniques, and show how to use *Maple* to analyze differential equations and their solutions with these techniques. As you encounter each new technique, you should ask: Under which of the six themes listed above does it fit? Answering that question will help you to organize the course material.

In order to take advantage of this modern approach, you must learn to use a computer platform effectively. We give detailed instructions for using *Maple* on three specific types of computers. In this way, we minimize the time required to learn to use the computer platform.

Finally, the software system we have chosen provides an additional valuable opportunity: it enables the production of "Worksheets" that combine textual commentary with the numerical, symbolic, and graphical output of the mathematical software system. Engineers and scientists have to develop not only skills in analyzing problems and interpreting solutions, but also the ability to present coherent conclusions in a logical and convincing style. You should use the Worksheet capabilities of the software to submit solutions to the computer assignments in such a style. This is excellent preparation for the professional requirements that lie ahead.

Student's Guide

The chapters of this book can be divided into three classes: descriptions of the computer platform, supplementary material on ordinary differential equations (ODE), and computer problem sets. Here is a brief description of the contents.

The computer platform chapters require no prior knowledge of computers. Chapter 2 explains how to start *Maple* in *Windows*, on the *Macintosh*, and in the *X Window System*. Chapter 3 introduces some basic *Maple* commands. You should work through Chapters 2 and 3 while sitting at your computer. Then you should work the problems in Problem Set A. These steps will bring you to a minimal level of competence in the use of *Maple*. You should complete these tasks within one week after the semester begins. After that, read Chapter 4, which contains detailed instructions for manipulating the *Maple* Worksheet interface on your computer. Finally, Chapter 8 is an introduction to the more sophisticated *Maple* commands that will be useful in later problem sets. It is best to read this chapter after working some problems in Problem Set B, but before starting on Problem Set C.

Keep in mind that the computer platform chapters are intended to teach you enough about *Maple* to study ordinary differential equations. You can explore *Maple* in more depth using its extensive online help resources. You can also consult any of a number of books about *Maple*. Guides to the *Maple* interface on various computers are available from Waterloo Maple Software.

The eight ODE chapters (5-7, 9-13) are intended to supplement the material in your text. The emphasis in this book differs from that found in a traditional ODE text. The main difference is less emphasis on the search for exact formula solutions, and greater emphasis on qualitative, graphical, and numerical analysis of the equations and their solutions. Furthermore, the commands for analyzing differential equations with *Maple* appear in these chapters.

The third collection of documents in this book consists of six sets of computer problems. The problem sets form an integral part of the course. Solving these problems will expose you to the qualitative, graphical, and numerical features of the course. Each set contains about fifteen problems; your instructor will decide which problems to assign, and when and how you are to submit them.

You can most profitably attack the problem sets if you plan to do them in two distinct sessions. Begin by reading the problems and thinking about the issues involved. Jot down some notes about how *Maple* can help you solve the problems. Although it is tempting to sit at the computer and start typing immediately, you will find that approach to be inefficient and potentially frustrating. Some preparation with pencil and paper will be useful. Then go to the computer and start solving the problems. If you get stuck, save your work and go on to the next problem. You should expect to spend at least one hour per problem. If there are things you'd like to discuss with your instructor, print out the relevant parts of the Worksheet to take with you. Then log out, go home, and try to digest what you've done—decide what's right, what's wrong, what could be done better, what surprises you encountered, *etc*. Talk to your instructor or your peers about anything you don't understand. This first session should be attempted a week before the assignment is due. After you have reviewed your output and obtained answers to your questions, you are ready for your second session. At this point, you should fill the gaps, correct your mistakes, and polish your Worksheet. This will take another hour or so per problem. Although you may find yourself spending extra time on the first few problems, if you read the *Maple* chapters carefully, and follow the suggestions above, you should steadily increase your level of competence in using *Maple*.

The end of this book contains two useful sections: a Glossary and a collection of Sample Worksheet Solutions. The Glossary contains a brief summary of some *Maple* commands, options, and built-in functions. The Sample Worksheet Solutions show how we solved several problems from this book. These samples will serve as guides when you prepare your own Worksheet solutions. Emulate them. Strive to prepare coherent, organized solutions. Combine *Maple*'s input, output, and graphics with your own textual commentary and analysis of the problem. Edit the final version of your solution to remove syntax errors and false starts. You will soon take pride in submitting complete, polished solutions to the problems.

Instructor's Guide

The philosophy that guided the writing of this book is explained at the beginning of this chapter. Here is a capsule summary of that philosophy. We seek:

- To guide students into a more interpretive mode of thinking.
- To use a mathematical software system to enhance students' ability to compute symbolic and numerical solutions, and to perform qualitative and graphical analysis of differential equations.
- To develop course material that reflects the current state of ODE and emphasizes the mathematical modeling of physical problems.
- To minimize the time required to learn to use the computer platform.

Our material consists of computer platform documents, ODE supplements, and computer problem sets. Here are our recommendations for integrating this material into the course.

Platform Documents. These are discussed in the preceding section. Proper use of these documents will quickly propel students to a reasonable level of proficiency. Students should read Chapters 2 and 3, and work Problem Set A within the first week of the semester. They should read Chapter 4 immediately after that, but they can delay reading Chapter 8 until after they start Problem Set B.

ODE Chapters. These eight chapters supplement the material in a traditional text. We use *Maple* to study differential equations using symbolic, numerical, graphical, and qualitative methods. We emphasize the following topics: stability, comparison methods, numerical methods, direction fields, and phase portraits. These topics are not emphasized to the same degree in a traditional text. We suggest that you incorporate this new emphasis into your class discussions. You should devote some class time to each chapter. Specific guidelines are difficult to prescribe, and the required time varies with each chapter, but on average we spend 30–40 minutes per chapter in class discussion. Of course, students will pay greater attention to the chapters if the material appears on quizzes and exams.

The structure of this book requires that numerical methods be discussed early in the course, immediately after the discussion of first order equations. The discussion of numerical methods should be directed toward the use of *Maple*'s numerical ODE solvers.

Computer Problem Sets. There are six computer problem sets. The topics addressed in the problem sets are:

- (A) Practice with *Maple*
- (B) First Order Equations
- (C) Numerical Solutions of Differential Equations
- (D) Second Order Equations
- (E) Series Solutions and Laplace Transforms
- (F) Systems of Equations

Problem Set A is a practice set designed to acquaint students with the basic symbolic and graphical capabilities of *Maple*, and to reacclimate them to calculus. It should be assigned on the first day of class, to be completed within one week. It is best to assign each remaining problem set at least two weeks before the due date. The course material corresponding to the problem set should be covered during those two weeks. It is good practice to set aside a small amount of class time to discuss questions that arise in students' first attempts to work the problems. That should typically be 3–5 days before the due date. (In the student guide, we suggested a two-stage plan of attack for completing the problem sets. Your class discussion should ideally occur after the first stage.)

We recommend that you assign all of Problem Set A and 3–6 problems from the other problem sets. Students should be required not only to analyze the problems critically, but also to present their analysis in coherent English and mathematics, displayed appropriately on their printouts. To accomplish this, students must master the editing features in the Worksheet interface of *Maple*. These features are discussed in the computer platform chapters. This skill is very important: Engineers and scientists do not just solve problems; they must also present their ideas in cogent and convincing fashion. We expect students to do the same in this course.

It is important to keep in mind that computer files can be copied effortlessly. You should choose a policy regarding student collaboration, and make this policy clear. For example, you might allow students to discuss the assignments, but require them to create the solution Worksheets themselves. Or you might allow students to work jointly and to submit collective solutions. We have found that allowing students to collaborate (in teams of no more than three) can be very beneficial. You should also vary the problems you assign from semester to semester.

Hardware Considerations. Before using these materials, you should make sure that your institution has enough machines with enough computing power to run *Maple*. Solving the problems in this book in a reasonable amount of time requires a computer with at least 16 megabytes of random access memory (RAM). On some platforms, *Maple* requires a numeric coprocessor (sometimes called a floating point unit, or FPU); it is recommended for all platforms.

Software Considerations. New versions of software appear frequently. When a complex system like *Maple* changes, many commands work better than they did before, some work differently, and a few may no longer work at all. As this book goes to press, the current version is *Maple V Release 4*; this is the version we describe. Most problems in the book can also be solved using *Release 2* or *Release 3*, though the syntax in earlier versions is sometimes substantially different. Many problems cannot be solved with a version older than *Release 2*.

Generalities. You do not have to be a *Maple* expert to use this course supplement. On the other hand, both you and your students will benefit if you spend some time working with *Maple*. We recommend that, in addition to working through Problem Set A, you work through the problems you assign in each set, preferably on the same computers that the students will use.

Our practice is to spend little class time discussing *Maple* per se. We try not to lose sight of the fact that we are teaching a course in *mathematics*, not *Maple*. We expect students to gain expertise in *Maple* from this book, from experimentation, from *Maple*'s extensive online help, from talking to each other, and from questions addressed to the instructor after class and in office hours.

Students often try to establish the following mode in their mathematics courses: "Give me a formula, I will give you a number!" However, we want students to understand the motives and methods behind a formula and its use. In the graphical environment offered by *Maple*, students sometimes try to establish a new mode: "Give me a program, I will give you a picture!" We ask not only that students produce a picture, but also that they interpret it, analyze the phenomenon it represents, and describe clearly and coherently the conclusions they derive from it.

Sample Syllabus.   In this section, we describe how this supplement might be used with two different elementary differential equations texts: **Elementary Differential Equations**, Sixth Edition, by Boyce & DiPrima, (Wiley, 1996) and **Differential Equations and Boundary Value Problems**, by Edwards and Penney, Prentice-Hall, 1996.

   This supplement contains enough material to address the topics in the first nine chapters of Boyce & DiPrima or the first eight chapters of Edwards & Penney. This material is more than ample for one semester or two quarters of study. If the course is given in a 15-week semester, then it is likely that portions of these chapters would have to be omitted. For example, at the University of Maryland at College Park, we omit Chapters 5 and 6 of Boyce & DiPrima. Because of the emphasis in this supplement on *Maple*'s numerical ODE solver, it is imperative that numerical methods be treated early in the course.

   The following schedules are for a typical semester. We assume a MWF schedule with 38 lectures and three days for exams).

Syllabus for use with Boyce & DiPrima

B & D Sections	Material from this book Chapters	Prob. Sets	No. of Lectures
2.1–2.8	1–6	A, B	10
8.1–8.4	7, 8	C	3
3.1–3.9	9	D	10
4.1–4.2	12		1
7.1–7.7	12	F	7
9.1–9.5	13	F	7

Syllabus for use with Edwards & Penney

E & P Sections	Material from this book Chapters	Prob. Sets	No. of Lectures
1.1–1.5	1–6	A, B	7
2.1–2.6	7, 8	C	6
3.1–3.6	9	D	8
4.1–4.3	12	F	4
5.1–5.3, 6.1–6.5	13	F	8
7.1–7.6	11	E	5

The following schedule indicates where exams and problem set due dates could be placed for the Boyce & DiPrima syllabus. Each problem set should be assigned at least two weeks in advance of the due date.

Item	Lecture Number
Problem Set A	5
Exam 1	11
Problem Set B	13
Problem Set C	20
Exam 2	25
Problem Set D	27
Exam 3	38
Problem Set F	40

Here is a sample grading scheme that gives ample weight to the *Maple* assignments: 3 Exams (@ 15% each) = 45%; 4 Computer Problem Sets (@ 5% each) = 20%; Quizzes or Standard Homework = 5%; and the Final Exam = 30%.

Instructors have a great deal of flexibility in deciding how to integrate the material from this supplement into a course. By choosing among the ODE chapters (Chapters 5–7 and 9–13), the instructor can decide which aspect of the course to emphasize. In order to emphasize numerical methods, you could stress Chapter 7 and Problem Set C. In order to emphasize symbolic computation, you could stress Chapters 5, 10, and 11 and Problem Set E. In order to emphasize geometric methods, you could stress Chapters 6, 9 and 13 and Problem Sets B and D. Chapter 12 incorporates both symbolic and geometric methods.

By adjusting the number of problems assigned from each problem set, the instructor can adjust the level of intensity. For example, some of our colleagues spread the problems out in a more uniform way during a semester, instead of assigning several substantial problem sets. Although we prefer to assign a few substantial problem sets, the material accommodates either mode.

The authors of this book maintain a Web site at

```
http://www.wam.umd.edu/~stuck/ODE
```

Chapter 2

Getting Started with *Maple*

In this chapter, we describe the basic skills needed to use *Maple* on several different computer platforms. In particular, we describe how to open, save, and print files. These are all features of the *interface* between *Maple* and the particular computer operating system that you are using. These instructions are distinct from the actual commands you will use to do mathematics with *Maple*. The next chapter introduces the mathematical aspects of *Maple*.

This chapter has three parts, one for each of the following user interfaces: *Windows 95*, *Macintosh* and the *X Window System*. You should find out what kind of computer and operating system you will be using, and then sit down at the computer with this guide and read the appropriate section. We note that *Windows 95* is most commonly found on PCs, and the *X Window System* on *UNIX* workstations. If you are using *Windows 95* or the *Macintosh* operating system on a networked computer, some of the features we describe below may be different for your particular installation. You may need to consult a local expert if you are using a different operating system (such as *NeXTStep*, *OS/2*, *Windows 3.1*, or *Windows NT*).

All of the systems we describe are window-based interfaces, which are manipulated using the keyboard and the mouse. Moving the mouse moves a small arrow (called a *pointer*) on the screen. To *click* on an object, place the pointer over the object and press the mouse button (if your mouse has more than one button, use the left button). To *double-click*, position the pointer and press the button twice in rapid succession. To *drag* an object, position the pointer over it, press and hold the mouse button, slide the object to where you want it, and then release the mouse button. A window can be dragged by its titlebar.

Maple in *Windows 95*

When you sit down at the computer, it may be turned off or it may be running a screensaver program. If it's off, turn it on and wait until it finishes booting. If a screensaver is running, move the computer's mouse to refresh the display. If all goes well, the screen should now contain various windows and/or icons.

In *Windows 95*, you manage programs, or *applications*, from the "Taskbar". The Taskbar usually appears as a slender grey rectangle across the bottom of the screen. At its left end, there is a button labelled "Start". Pressing the "Start" button brings up a menu of choices, including **Help**, **Settings**, and **Programs**. Clicking on one of these choices either opens a submenu with more choices (submenus may be nested several layers deep) or launches an application.

In addition to the "Start" button, the Taskbar may contain other buttons. Each button represents an active application. When you start an application, a button is added to the Taskbar; clicking on the button will reactivate the application.

In addition to the Taskbar, the *Windows 95* desktop contains several icons. Double-clicking on the icon labelled "My Computer" will open a window containing icons that represent the disk drives and other hardware installed in the computer. Double-clicking on a drive icon opens a window showing the files and folders stored on the drive. Double-clicking on a file will start the application associated with the file, allowing you to view and modify the file's contents. Double-clicking on a folder will open a window displaying its files and subfolders.

If the computer is networked, then there will be an icon labelled "Network Neighborhood". Double-clicking on the "Network Neighborhood" icon will open a window showing the resources available on the network. Double-clicking on these icons may allow you to manipulate the files and folders stored on other computers on the network.

If a window is partially obscured, you can bring it to the front by clicking on any part of it, or by clicking on the corresponding button on the Taskbar. In the upper right-hand corner of a window are three small boxes. Clicking the box marked with an "X" will quit the application. Clicking the box marked with a rectangle will expand the window to fill the entire screen. Clicking the box marked with a small line will make the window disappear; you can make it reappear by clicking on the corresponding button on the Taskbar.

Online Help for *Windows 95*.   If you have never used *Windows 95* before, it's probably a good idea to take the "Windows Tour". To do this, press the "Start" button, and then choose **Help**. This will open a help window. Click the tab labelled "Contents", then double-click on "Tour: Ten minutes to using Windows". The tour covers the basics of managing windows and applications.

Starting *Maple*.   The *Maple* program uses a lot of memory, so you should close other applications before starting *Maple*. An application can be closed by choosing the **Quit** or **Exit** button in the **File** menu belonging to the application, or by clicking on the small box labelled with an "X" in the upper right-hand corner of the window.

Once you have closed the other applications, you are ready to start *Maple*. Press the "Start" button and choose **Programs**. The icon for the *Maple* application is a maple leaf with the name "Maple V Release 4" next to it. The location of this application will depend on your particular machine, so you may have to hunt around to locate it. Try looking in submenus that have the word "Application" or "Math" in their titles. When you've found the correct menu item, click on it. After a brief delay, a new window labeled "Maple V Release 4" will appear on the screen. This window is the main window for the *Maple* program. It will contain some menus and buttons at the top, and will also contain a subwindow (or *child window*) labeled "Untitled (1)". This child window is a *Maple* "Worksheet". You can now start typing in the Worksheet.

Saving Your Worksheet on a Diskette. To save your Worksheet onto a diskette, you'll need to buy some diskettes for the computer you are using (usually 3.5 inch, 1.44 megabyte High Density (HD) diskettes). These are inexpensive and available in many book stores and computer stores.

Insert a diskette in the slot in the front of the computer. To find out what files are stored on the diskette, move or hide windows until you can see the "My Computer" icon on the *Windows 95* desktop. Double-click on the "My Computer" icon. Move the mouse pointer to the icon representing the floppy drive. (The icon will be labelled "Floppy (A:)" or "Floppy (B:)". Make sure that you are pointing to the correct floppy drive.) Double-click on the floppy icon to view the files it contains.

If the disk is unformatted, or formatted for a different kind of computer, you will have to format it before you can use it. *Caution:* Formatting a disk removes all information from it. Follow the instructions above to make certain that you do not have important files on a disk before formatting it.

To format a diskette, double-click the "My Computer" icon and point to the correct floppy drive. Press the right-hand mouse button to open a menu, and click on **Format**.... A dialog panel will appear. In this panel, you can give a name to the diskette by typing in the "Label" box. To format an unformatted diskette, choose "Full" as the format type. (A "Quick" format will only work on previously formatted diskettes.) Once you have selected the proper format type and typed in a name, click on the "Start" button in the dialog panel. A full format may take a few minutes.

Once you have a properly formatted diskette inserted in the drive, make sure that your *Maple* window is active by clicking in it. A darkened titlebar indicates that the window is active. Now select **Save As**... from the **File** menu. A dialog panel will appear on the screen. Near the bottom of the panel is a box labeled "Drives:". The drive that has your diskette will be either "A:" or "B:". Select the proper drive by clicking on its name. (If you don't see the name of the drive, use the arrow button to move around in the list of drives.) Once you have the proper drive selected, click in the box labeled "File Name:" (near the upper left-hand corner) and then type the name you'd like to use for the Worksheet. The name should have the extension ".mws"; for example, "Problem1.mws". Then click "OK". If you get an error message, it is likely that you've inserted the disk incorrectly or in the wrong drive, or the disk is not formatted correctly. If all goes well, a copy of the file will be saved on the diskette, and the name of the *Maple* window will change from "Untitled (1)" to the name you typed. To verify that the file has been copied, you can use the drive icon in the "My Computer" window.

Once a Worksheet has a title, you can save further changes by simply selecting **Save** from the **File** menu. *Maple* automatically keeps track of the filename. If you want to save the Worksheet under a different name, you'll have to use the **Save As**... button again. You should save your work frequently.

If the computer you are using is networked, and you have an account on the network, you may be able to save your Worksheet on a central file server. Consult your system administrator for instructions.

Opening a Previously Saved Worksheet.   To open a saved Worksheet, click on the **File** button, and then click on the **Open** button. A new dialog panel will appear. This panel is essentially identical to the one that appeared as a consequence of selecting the **Save As**... button. You can navigate in this panel by clicking on drives and/or directories until the file you want appears in the box labeled "File Name:". After selecting the file you want, click on the "OK" button. *Maple* worksheets have the filename extension ".mws".

To open a new Worksheet, click the **New** button in the **File** submenu. You can open and save as many Worksheets as you like, using a different name for each one, as long as you don't use up the available disk space.

If you have several Worksheets open, you may want to close one or more. To close a Worksheet, make it active by clicking in it or by selecting its name from the menu labeled **Window**, and then select **Close** from the **File** menu. *Maple* will prompt you to save the Worksheet if the Worksheet has changed since the last save. Do not attempt to close a Worksheet by selecting **Exit** from the **File** menu; that will terminate your *Maple* session entirely.

Printing Your Worksheet.   To print a Worksheet, first make sure the Worksheet you want to print is the active window. Then select **Print**... from the **File** menu. A new panel will appear offering various options. One option is to print a selected range of pages. You can do this by typing the appropriate page numbers in the "From" and "To" boxes. When you're ready to print, click on the "OK" button.

Interrupting Calculations.   In the "Tool Bar" near the top of the Worksheet is a button with a stop sign on it. If *Maple* is hung up in a calculation, you can usually stop it by clicking this button.

Ending the *Maple* Session.   It is important to quit *Maple* when you are done working. First make sure that you have saved your work. Then select the **Exit** button from the **File** menu. If you haven't saved all your work, *Maple* will give you one last chance to save it. When you have exited *Maple*, its menu and any open Worksheets will disappear. Be sure to quit any other programs you have been using before you leave the machine.

Don't forget to eject your diskette before you leave. On most machines, you eject the diskette by pushing the button just below the slot for the diskette.

Maple on a *Macintosh*

When you sit down at a *Macintosh* it may be turned off, or it may be running a screensaver program. If it's off, turn it on and wait until it is finished booting. If a screensaver is running, move the computer's mouse to refresh the display. If all goes well, along the top of the screen you should see the following (from left to right): an Apple icon; a series of menu buttons; the time; a "Help" icon indicated by a question mark; and in the upper right-hand corner, an "Application" icon. Click and hold on the Application icon. A menu will appear that will show, among other things, a list of applications running on the *Macintosh*. There will be a check mark on the button corresponding to the current application. Drag the mouse down to the "Finder" button, and then release. Your screen may change, depending on which application was active, and you should see one or more windows or panels on the screen. You are now in the Finder.

The Finder is the *Macintosh* program for finding, copying, moving, and organizing files. Each of the icons in the Finder represents either a file or a folder; a folder is a group of files bundled together. Clicking on a file or folder icon will *select* the icon, and the icon will be darkened. Double-clicking on a folder icon will cause the Finder to bring up a new window displaying a set of icons representing the files in that folder. Double-clicking on a file icon will open the file. The *Macintosh* tries to open a file using an appropriate *application*, or program. For example, *Maple* files will be opened using the *Maple* program. Double-clicking on such a file will open it in *Maple*. You can usually tell the type of a file by looking at its icon. For example, the icon for *Maple* files is a maple leaf.

If a window is partially obscured you can bring it to the front by clicking on any part of it. Clicking in the small box in the upper left-hand corner of a window will make the window disappear.

Moving files from one folder to another is done as follows: find the folder into which you want to move (or copy) a file. This folder can be either in iconic form, or in the form of an open window showing its contents. Then find the file you want to move and drag it to the target folder. An icon representing the file will appear in the target folder. (The file will be moved or copied, depending on who owns the file and folder.) To delete a file, drag its icon into the trash can at the bottom right-hand corner of the screen. Then select **Empty Trash**... from the **Special** menu at the top of the screen. A warning panel will appear on the screen asking you to confirm that you want to delete the files in the trash can.

Online Help for the *Macintosh*. If you have never used a *Macintosh* computer before, it is probably a good idea to look at the *Macintosh* "Tutorial" or the "Macintosh Guide", both of which are accessible from the **Help** menu. The tutorial will cover the basics of managing files and opening applications and will help you become comfortable using the *Macintosh*. You can also get a rudimentary kind of help by turning on the **Show Balloons** option in the **Help** menu.

Starting *Maple*.   The *Maple* program uses a lot of memory, so before you start *Maple* you should quit any other applications that are running on the *Macintosh*. To do this, use the menu attached to the Application icon in the upper right-hand corner of the screen.

The icon for the *Maple* application is a maple leaf with the name "Maple" printed below. It should be in a folder called "Maple", or something similar. The location of this folder will depend on your particular machine, so you may have to hunt around in the Finder to locate it. Try looking in folders that have the word "Application" or "Math" in their titles. When you've found the icon, double-click on it to open it. After a brief delay, a new window labeled "Untitled (1)" will appear on the screen, and the menu buttons along the top of the screen will change. You can now start typing in the Worksheet.

Saving Your Worksheet on a Diskette.   To save your Worksheet onto a diskette, you'll need to buy some diskettes for the computer you are using (usually 3.5 inch, 1.44 megabyte High Density (HD) diskettes). These are inexpensive and available in many book stores and computer stores. Insert a diskette in the slot in the front of the *Macintosh*. If the disk is unformatted, or formatted for a different kind of computer, a panel will appear asking you whether you want to format it. If you click **Initialize**, a panel will appear asking whether you really want to erase the diskette. If you click **Erase**, the *Macintosh* will prompt you for a name for the diskette. Type in a name, then hit RETURN. It will take a few minutes to initialize the diskette.

Once you have a properly formatted diskette inserted in the drive, an icon representing the diskette will appear along the right-hand edge of the screen. Make sure that your *Maple* window is active by clicking the mouse in it. A darkened titlebar indicates that the window is active. Now select **Save As**... from the **File** menu. A new panel will appear on the screen. Select the button **Desktop** on the right-hand side. In the center of the panel you should see the name of your diskette. Click on the name, and then click **Open**. The panel will change, and there will be a box labeled "Save as:". Type the name you'd like to use for the Worksheet ("Assignment 1", for example), and then click **Save**. A copy of the file will be saved on the diskette, and the name of the main window will change from "Untitled (1)" to the name you typed. To check that the file has been copied, double-click on the diskette icon. A new window will appear containing a list of files on the diskette. To continue working with *Maple*, click on your Worksheet.

Once a Worksheet has a title, you can save further changes by simply selecting **Save** from the **File** menu. *Maple* automatically keeps track of where you want to save it. If the diskette on which you originally saved the file is not in the drive, the *Macintosh* will pop up a panel with an error message, or prompt you to insert the diskette. In the latter case, you can get rid of the panel by typing "COMMAND-.". (The COMMAND keys are located on either side of the space bar, with both an apple and a cloverleaf-like symbol.) If you want to save the Worksheet under a different name, you'll have to use the **Save As**... button again. You should save your work frequently.

If the computer you are using is networked, and you have an account on the network, you may be able to save your Worksheet on a central file server. Consult your system administrator for instructions.

Opening a Previously Saved Worksheet.   To open a saved Worksheet, click on the **File** button, and then click on the **Open** button. A new dialog panel will appear. This panel is essentially identical to the one that appeared as a consequence of selecting the **Save As**... button. You can navigate in this panel by clicking on drives and/or directories until you find the file you want. After selecting the file, click on the "Open" button.

To open a new Worksheet, click the **New** button in the **File** submenu. You can open and save as many Worksheets as you like, using a different name for each one, as long as you don't use up the available disk space.

If you have several Worksheets open, you may want to close one or more. To close a Worksheet, make it active by clicking in it or by selecting its name from the menu labeled **Window**, and then select **Close** from the **File** menu. *Maple* will prompt you to save the Worksheet if the Worksheet has changed since the last save. Do not attempt to close a Worksheet by selecting **Exit** from the **File** menu; that will terminate your *Maple* session entirely.

Printing Your Worksheet.   To print a Worksheet, first make sure the Worksheet you want to print is the active window. Then select **Print**... from the **File** menu. A new panel will appear offering various options. One option is to print a selected range of pages. You can do this by typing the appropriate page numbers in the "From" and "To" boxes. When you're ready to print, click on the **Print** button.

Interrupting Calculations.   In the "Tool Bar" near the top of the Worksheet is a button with a stop sign on it. If *Maple* is hung up in a calculation, you can usually stop it by clicking this button. You may also be able to stop it by pressing "COMMAND-.".

Ending the *Maple* Session.   It is important to quit *Maple* when you are done working. First make sure that you have saved your work. Then select the **Exit** button from the **File** menu. If you haven't saved all your work, *Maple* will give you one last chance to save it. When you have exited *Maple*, its menu and any open Worksheets will disappear. Be sure to quit any other programs you have been using before you leave the machine. (You can check to see which programs are running by using the Application menu in the upper right-hand corner of the screen. You do not have to quit the Finder.)

Don't forget to eject your diskette before you leave. You can eject the diskette by selecting **Eject Disk** from the **Special** menu in the Finder, or by dragging the diskette icon into the trash can.

Maple in the *X Window System*

We assume that you are using a computer running some version of the *UNIX* operating system and that you are using the *X Window System*. We also assume that you have an account on the computer and know how to log on. The *X Window System* is highly customizable, and you will have to learn the idiosyncrasies of your particular installation. In many installations, windows or panels will appear with a titlebar at the top, with buttons in the right- and left-hand corners. You can bring a partially obscured window to the front by clicking its titlebar. To do much more than this with your windows, you will have to consult site-specific documentation and/or your system manager.

One of the windows on your screen should be a terminal window (probably titled "xterm" or "login"). You can type *UNIX* commands in this window. Here is a list of the basic *UNIX* commands for locating, moving, copying, and deleting files.

UNIX Command	Effect
ls	lists your files
cp file.a file.b	puts a copy of file.a into file.b
mv file.a file.b	renames file.a to file.b
rm file.a	deletes file.a

In fact, you should be able to do most of the necessary file manipulation of your *Maple* files from within the *Maple* program (to be explained below), but on occasion you may need to maneuver your *Maple* files exactly as you would any other files in the *UNIX* environment.

Online Help for the *X Window System*. The usual way to obtain help in a *UNIX* environment is to consult the manual pages using the "man" command. For example, by typing the command "man ls" you can learn about the features of the "ls" command, including options that can be used with it. If you are a *UNIX* novice, you may find it instructive to try the "man" command on some of the basic system commands listed above.

Starting *Maple*. *Maple* is often started from an xterm window by typing a special command at the *UNIX* prompt. The command may be "maple" or "xmaple". Alternatively, *Maple* might be started on your system by clicking on a special menu button. You may need to ask your system administrator for the precise command or start-up procedure. Whatever the procedure, a new window labeled "Maple V Release 4" will appear on the screen. This window is the main window for the *Maple* program. It will contain some menus and buttons at the top, and will also contain a subwindow (or *child window*) labeled "Untitled (1)". This child window is a *Maple* "Worksheet". You can now start typing in the Worksheet.

Saving Your Worksheet. To save a Worksheet, click on the **File** button in the menu row. A new menu will appear below the button. Click on the **Save As**...

button. A panel will appear in the middle of the screen. By double-clicking on a directory name in the Directory region, you cause the selected directory to become the current directory; the Directory and File regions will change to show the subdirectories and files in the new current directory. If there are more subdirectories in the current directory than will fit in the Directory region, you can use the mouse to manipulate the slider to the right of the Directory region to scroll up and down through the list of subdirectories.

Once you have reached the directory in which you wish to save the Worksheet (by double-clicking on successive subdirectories or typing a pathname into the Filename box) you should place the pointer at the end of the pathname in the selection box and click. Then type in the name you wish to give to your Worksheet. The name should end with the suffix ".mws". Then click on the **Save** box in the menu row. The panel will disappear, the Worksheet will be saved, and its title will change to the name you typed. If after working some more you want to save your Worksheet again using the same name, all you have to do is click on the **Save** button instead of the **Save As**... button. If you want to save it under a new name, you must repeat the steps above. You should save your work frequently.

Opening a Previously Saved Worksheet. To open a saved Worksheet, click on the **File** button, and then click on the **Open** button. A panel that is essentially identical to the one that resulted from selecting the **Save As**... button will appear. You can navigate the directory tree by double-clicking on directories (or by typing a pathname into the Filename region) until the file you want appears in the File region of the panel. Then double-click on the name of the appropriate file (or click on the name and select "OK".

To open a new Worksheet, click the **New** button in the **File** menu. You can open and save as many Worksheets as you like, using a different name for each one, as long as you don't use up the available disk space.

If you have several Worksheets open, you may want to close one or more. To close a Worksheet, make it active by clicking in it, and then select **Close** from the **File** menu. *Maple* will prompt you to save the Worksheet if the Worksheet has changed since the last save. Do not attempt to close a Worksheet by selecting **Exit** from the **File** menu; that will terminate your *Maple* session entirely.

Printing Your Worksheet. To print a Worksheet, select **Print** from the **File** menu. A panel will appear in the center of the screen. Near the top of the panel are two small buttons labeled "Output to File:" and "Print Command:". If you click on the button labeled "Print Command:", your Worksheet will be directed to the default printer. If you click on the button labeled "Output to File:", the Worksheet will be written to a PostScript file with suffix ".ps". The default filename will be shown in the box to the right of the label. If you want to send the output to a PostScript file with a different filename, you can click in this box and type in a new filename. Finally, when you have everything as you want it, click on the "Print" button at the bottom of the panel.

Interrupting Calculations.   Near the top of the Worksheet is a button with a stop sign on it. If *Maple* is hung up in a calculation, you can usually stop it by clicking on this button.

Ending the *Maple* Session.   When you are done working, it is important to quit *Maple*. First make sure that you have saved your work. Then select the **Exit** button from the **File** menu. If you haven't saved all your work, *Maple* will give you one last chance to save it. Be sure to quit any other programs you have been using before you log out and leave the machine.

Chapter 3

Doing Mathematics with *Maple*

These instructions are designed to get you started with the *Maple* software. Before you read this chapter you should already have read the appropriate section of the preceding chapter, *Getting Started with Maple*. After you have read that section, you will know how to get *Maple* up and running. Then you should read this chapter, trying out the commands in a *Maple* Worksheet as you go along. For further practice with these *Maple* commands, work the problems in Problem Set A.

Input and Output. You interact with *Maple* using Worksheets. On your computer screen a Worksheet appears as a window with various kinds of text and graphics in it. But *Maple* also treats a Worksheet as a *document*, and in particular *Maple* knows how to break the Worksheet into pages to produce a finished, printed version of the Worksheet. A Worksheet is divided into *regions* of different types, containing input, output, graphics and text. We'll discuss the structure of Worksheets in detail later, but in this chapter we focus on input regions and output regions. An input region (or *Execution Group*) is a place where you type commands for *Maple* to evaluate, and is marked with a "greater than" prompt (>). An output region is a place where *Maple* places its response to an input.

To enter a command in *Maple*, you type the command in an input region, put a semicolon at the end of the command, and then press the ENTER key. (on the *Macintosh*, the RETURN key) If you type ENTER without first typing a semicolon, *Maple* will complain If this happens, backup, type a semicolon and hit ENTER again. If you want a simple linefeed in an input region, type SHIFT-ENTER (SHIFT-RETURN on the *Macintosh*). This is useful when typing a long formula, or to group several commands or expressions together in a single input region. You can enter several commands in a single input region as long as you end each command with a semicolon.

As an example, try typing **1 + 1;** (with the semicolon). It should appear on the screen near the top. You have given *Maple* an expression to evaluate. To find out the answer, press ENTER. *Maple* will display the answer. Your Worksheet should now look something like this:

> **1 + 1;**

$$2$$

The symbol ">" is the input prompt in *Maple*. The output of the command (in this case "2") is in an output region. Material in the output regions is specially formatted in mathematical notation.

Arithmetic.   As we have just seen, *Maple* can do arithmetic like a calculator. You can add with "+", subtract with "−", multiply with "*", divide with "/", and exponentiate with "^". For example:

> **2*5 + 3^2 − 4/2;**

$$17$$

> **(5 + 3)*(5 − 3);**

$$16$$

Maple requires you to use the "*" symbol explicitly to express multiplication. Typing "2 5" instead of "2*5", or "2x" instead of "2*x", will cause a *syntax error*—*Maple*'s way of letting you know that it didn't like the way you typed the expression.

Maple differs from a calculator in that it often does exact arithmetic rather than using decimal approximations. For example, if you type in **4/7**, *Maple* will simply respond $\frac{4}{7}$. To force *Maple* to give you a decimal approximation, you must use the built-in *Maple* command **evalf** (which means "evaluate as a floating-point number").

> **evalf(4/7);**

$$.5714285714$$

To get 20 digits rather than 10 in the answer, you would type **evalf(4/7, 20)**. Another way to force *Maple* to use decimal approximations rather than exact numbers is to use a decimal point. For example, in *Maple*, **3.0** is treated differently from **3** (enter the fractions **1/3** and **1/3.0** in *Maple* and see what happens). It is important to remember this distinction because approximate, or floating point, arithmetic is usually faster than exact arithmetic. Floating point output is also quite different in appearance. To see an example of this, evaluate the following two commands in your Worksheet.

> **solve(x^3 − x − 3 = 0, x);**
> **solve(x^3 − x − 3.0 = 0, x);**

Errors in Input.   If you make an error in an input line and *Maple* can't understand you, it will print an error message. For example:

> **5 + 3x − 2 ;**
```
Syntax error, missing operator or ';'
```

The error here is that there is a missing multiplication operator "*". Note that *Maple* places the cursor at the place where it first notices the error; however, the actual error may have occurred earlier. *Maple* may be able to correct the syntax error automatically if you click on the "syntax auto-corrector", the button in the **Context Bar** at the top of the worksheet with a check-mark inside a pair of parentheses. Missing multiplication operators, brackets and parentheses are the most common errors on input lines.

If you have made a mistake on an input line, you can edit the input line by using the mouse to position the pointer at the desired insertion point, clicking the button on the mouse, then typing (or deleting with the delete key), and finally pressing ENTER to reenter the line. When you reenter the line, *Maple* will write over the old output in the output region directly below.

Online Help. There are several ways to get online help in *Maple*. To get help on a particular topic, enter **?<topic>**. For example, the command **?evalf** will open a window that contains help on the **evalf** command. You can also access specific parts of the online help by typing **??** or **???** before a command name. Two question marks produce the syntax of the command, while three question marks take you directly to the examples.

You can also browse through the online help by pressing the **Help** menu button, and selecting **Contents**. You might also try the *Maple* "New User's Tour", which is located in the Worksheet called "Newuser.mws" in the "Examples" folder in the "MapleVR4" folder.

Many *Maple* commands have optional arguments to modify the behavior of the command. These optional arguments are described in the online help. We will use optional arguments extensively with plotting commands, and with numerical solutions of differential equations.

Algebra. You can add, subtract, multiply, and divide variables as well as numbers. Consider the following series of commands:

> **(x + y)*(x + y)^2;**

$$(x + y)^3$$

> **expand((x + y)^3);**

$$x^3 + 3x^2y + 3xy^2 + y^3$$

> **factor(x^3 + 3*x^2*y + 3*x*y^2 + y^3);**

$$(x + y)^3$$

Maple often makes minor simplifications to the formulas you type, but does not make any big changes unless you tell it to. The **expand** command forced *Maple* to expand the formula, and the **factor** command told *Maple* to put it back in factored form.

Maple also has a **simplify** command that tries to express a formula as simply as possible; sometimes this involves factoring, and sometimes multiplying things out. For example:

> **simplify((x^2 − y^2)/(x − y));**

$$x + y$$

> **simplify(exp(a + ln(b*exp(c))));**

$$be^{(a+c)}$$

Equations and Assignments.   In *Maple*, you use the combination "**:=**" (colon equals) to assign values to a variable. For instance,

> **x := 5;**

$$x := 5$$

will give the variable **x** the value 5 from now on. Whenever *Maple* sees an **x** it will substitute the value 5.

> **x^2 + 3*x*y + y;**

$$25 + 16y$$

To clear a variable, type **x := 'x'** . This reassigns to **x** its name, **x**.

> **x := 'x';**

$$x := x$$

> **x^2 + 3*x*y + y;**

$$x^2 + y + 3xy$$

Assignments can be quite general.

> **z := x^2 + 3*x*y + y;**

$$z := x^2 + 3xy + y$$

> **z + 7*y;**

$$x^2 + 3xy + 8y$$

> **y := 5;**

$$y := 5$$

> **z;**

$$x^2 + 15x + 5$$

Important: A common source of puzzling errors in a *Maple* session is to forget that you have defined variables. (*Maple* never forgets.) You can check the current definition of a variable by typing the variable name (followed by a semicolon) by itself on an input line. For example, to find the current value of the variable **z**, type **z** (followed by a semicolon). If *Maple* returns z, it means that **z** has not been assigned a value. If it returns anything else, **z** has been assigned a value.

An equation in *Maple* is indicated by an equals sign (without a colon). For example, to solve the equation $x + \sin y = 3$ for y, we first clear the variables x and y (to which we have assigned values) by typing **x := 'x'** and **y := 'y'**, and then type
> **solve(x + sin(y) = 3, y);**

$$- \arcsin(x - 3)$$

Built-in Commands and Functions. The commands **evalf**, **expand**, **factor**, and **solve** are examples of *Maple*'s built-in commands. *Maple* also has many built-in functions, including familiar mathematical functions like **sqrt**, **cos**, or **log**. *Maple* also has a few built-in constants, such as **Pi** (the number π), **I** (the complex number $i = \sqrt{-1}$) and **infinity** (∞).

Some examples:
> **log(2.0);**

$$.6931471806$$

Notice that **log** and **ln** are both names for the natural logarithm function, called "ln" in many texts.
> **sin(Pi/3);**

$$\frac{1}{2}\sqrt{3}$$

(To get a numerical answer one would use the **evalf** command.)
> **fsolve(exp(−x) = x, x);**

$$.5671432904$$

We have not described the command **fsolve**, but you can find out about it by entering **?fsolve** in your *Maple* Worksheet. We will explore many other *Maple* commands and functions in this book, including commands that can differentiate, integrate, or solve differential equations.

User-defined Functions and Expressions. It is also possible to define new functions in *Maple*. In the following example a polynomial function is defined and then evaluated at one point:
> **f := x −> x^3 + 7*x − 5;**

$$f := x \to x^3 + 7x - 5$$

> **f(2);**

In the definition of a function, the arrow is typed as a minus sign "−" followed by a greater than sign ">". The arrow represents a *transformation rule*: the function **f** takes a variable or value x and *transforms* it to $x^3 + 7x − 5$.

In *Maple*, it is important to distinguish a *function* from a simple *expression*. For example, **x^3 + 7*x − 5** is an expression. We can assign a name to an expression, as follows:

> **g := x^3 + 7*x − 5;**

$$g := x^3 + 7x − 5$$

Although this *expression* looks very much like the *function* we defined above, it is in fact quite different. The function is a transformation rule, while the expression is merely a quantity involving x. In particular, typing **g(2)** will not give the value of the expression at $x = 2$. (Try it to see what happens!) To evaluate an expression at a particular point, we must use the substitution command, **subs**.

> **subs(x = 2, g);**

$$17$$

This command means "substitute the value 2 for x in the expression g ".

Graphics. The basic plotting command in *Maple* is **plot**. It can be used to plot both functions and expressions.

> **plot(x^3 − x, x = −1..1);**

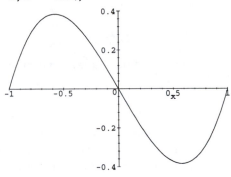

Notice that in this command we have plotted an expression, as opposed to a function. The syntax for plotting a function is similar:

> **plot(f(x), x = −1..1);**

An alternative syntax for plotting functions is **plot(f, −1..1)**. In these commands, **−1..1** indicates that the function or expression is to be plotted on the interval $−1 \leq x \leq 1$.

Plots can be misleading at times because the scale of the plot may not be what you expect and the origin may not be where you expect it to be. You should pay careful attention to the labeling of the axes. To get a plot with a specified vertical range, give the **plot** command a second range, as follows:

> **plot(x^3 − x, x = −1..1, y = −2..2);**

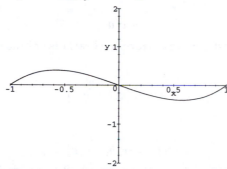

If you plan to print the Worksheet on a black-and-white printer, you should use the option **color = black** with the **plot** command.

Sequences, Sets and Lists. The three most common data structures in *Maple* are *sequences*, *sets*, and *lists*. A *sequence* is a collection of *Maple* objects separated by commas, as in $1, 2, 3$ or a, b, c. A *list* is a sequence surrounded by square brackets, as in $[1, 2, 3]$ or $[a, b, c]$. A *set* is a collection of *Maple* objects separated by commas and surrounded by braces, as in $\{1, 2, 3\}$ or $\{a, b, c\}$. These three data types are similar except for one important difference: In a set, order and repetition are irrelevant. Thus the sets $\{c, b, a\}$ and $\{a, a, b, c\}$ are identical to $\{a, b, c\}$. On the other hand, the lists $[c, b, a]$, $[a, a, b, c]$, and $[a, b, c]$ are all different. They are also different as sequences.

Of these three types of objects, the fundamental one is the sequence. A list is simply a sequence surrounded by brackets, and a set is a sequence surrounded by braces in which order and repetition are ignored.

You can create and name a sequence very easily:

> **s := b, c, d;**

$$s := b, c, d$$

You can also add items to the beginning or end of a sequence:

> **s := a, s, e;**

$$s := a, b, c, d, e$$

You can recover the individual elements of a sequence using brackets. Inside the brackets you put a number or range indicating the positions of the elements you wish to extract.

> **s[2];**

$$b$$

> s[2..4];

$$b, c, d$$

Lists and sets are constructed by wrapping brackets or braces around sequences.

> list1 := [s];

$$list1 := [a, b, c, d, e]$$

> set1 := {s};

$$set1 := \{c, b, a, d, e\}$$

Note that the order of elements is not preserved when we define a set.

Although it is easy to define a sequence, list, or set by simply typing in the elements, it is often convenient to automate this process. This can be done using either the **seq** command or the **$** (dollar sign) operator. To illustrate these facilities, we'll build a sequence in two different ways by evaluating a function $f(x)$ at the integers $0, \ldots, 4$.

> f := x -> x^2 - 1:
> f(i) $ i = 0..4;

$$-1, 0, 3, 8, 15$$

The symbol **$** is called the *sequence operator*. We can also define this sequence using the **seq** command.

> seq(f(i), i = 0..4);

$$-1, 0, 3, 8, 15$$

Both the sequence operator and the **seq** command increment the index by 1. More specifically, *Maple* initially sets the index equal to the first value in the specified range, and then increments it successively by 1 until the index becomes greater than the second value in the range, at which point *Maple* stops.

> {i $ i = 2.1..4};

$$\{2.1, 3.1\}$$

> [seq(i, i = -9..(2.1))];

$$[-9, -8, -7, -6, -5, -4, -3, -2, -1, 0, 1, 2]$$

You may want to build a sequence with a parameter in increments other than 1. To do this you just "rescale" the parameter. Here is a list of values of the exponential function at the points $0, 0.2, 0.4, \ldots, 1.0$.

> expvalues := [exp(0.2*i) $ i = 0..5];

$$expvalues :=$$
$$[1, 1.221402758, 1.491824698, 1.822118800, 2.225540928, 2.718281828]$$

Note that the parameter varies from 0 to 5 in increments of 1, but the argument of the exponential function, which is 0.2 times the parameter, varies from 0 to 1 in increments of 0.2.

Referring to Previous Output.   *Maple* automatically remembers the output of the last three statements you entered. A double quote (") refers to the output of the previous command. Two double quotes ("") refer to the second-to-last output, and three double quotes (""") to the third-to-last output. For example:

> `2 + 2;`

$$4$$

> `"^2;`

$$16$$

> `"""*5;`

$$20$$

A safer and more effective way to refer to past output is to assign the output to a variable. For example, we could type:

> `a := 2 + 2;`

$$4$$

> `a^2;`

$$16$$

> `a*5;`

$$20$$

This is more effective because it allows you to refer to the output by name anywhere in the Worksheet. It is safer because the output in the preceding region is not necessarily the most recent output generated by *Maple*. In fact, there is no way to tell by looking at the Worksheet which output is the most recent.

The Restart Command.   A simple and effective way to clear all the assigned values in a *Maple* session is to use the **restart** command. This command shuts down *Maple*'s computational engine, clears everything from memory, and then restarts the computational engine. Nothing will change on the screen. After you have typed **restart**, you can continue working. But keep in mind that all the variable assignments have been cleared, so you will have to evaluate the appropriate input regions to restore any assignments you want to use. We recommend that you use the **restart** command frequently. In particular, you should put the **restart** command at the top of every worksheet so that *Maple* starts with a clean slate each time you evaluate the Worksheet from the beginning.

Aborting Calculations.   If *Maple* gets hung up in a calculation, or seems to be taking too long, you can usually abort it by selecting the **Interrupt** or **Stop** button. It may take a while for *Maple* to abort the calculation. It's a good idea when using any computer program to save your work often, for an unexpected crash or power outage can wipe out all your unsaved work.

Visible State vs. Internal State.   When *Maple* saves a Worksheet, it saves the *visible state* of the Worksheet, *i.e.*, it saves exactly what you see on the screen and nothing else. If you open your Worksheet in a new *Maple* session, the Worksheet will contain the input and output statements, but the actual results will not be available to *Maple*'s internal computation engine. In order to use the results of your previous work in your current calculation, you must redo the old calculations. You can do this by stepping through the Worksheet, evaluating each input region. (There is also a menu button to evaluate the entire worksheet.)

Packages.   The *Maple* system actually consists of three parts. The part that reads input, displays output and manipulates Worksheets is called the *interface*. The other two parts of *Maple*—the *kernel* and the *library*—execute the commands you type. The kernel contains the core parts of *Maple* that are always available. The library contains additional commands that are only loaded into memory as needed. Most commands are loaded automatically without any intervention on your part. You don't need to concern yourself with these; for all practical purposes, you can pretend that they are part of the kernel.

To use certain *Maple* commands, however, you must tell *Maple* explicitly to load them. If you use one of these commands without explicitly loading it, *Maple* will not know the command and will simply echo the input line in the output area. For example,

> **contourplot(x^2 + y^2, x = −5..5, y = −5..5);**

$$\text{contourplot}(x^2 + y^2, x = -5..5, y = -5..5)$$

The command **contourplot** is a library command and must be loaded explicitly.

Most of the library routines that are not loaded automatically are bundled in *packages*. The two packages you will need for most of the problem sets are called **plots** and **DEtools**. The former contains a collection of specialized plotting routines including **contourplot**, **implicitplot**, and **display**. The latter contains additional routines such as **dfieldplot** and **DEplot** for studying differential equations. You can load all the commands in a package with a single command. For example, the command

> **with(plots):**

loads all the routines from the **plots** package. (If you end the command with a semicolon instead of a colon, then *Maple* prints a list of the commands contained in the package.) You can conserve memory by loading a single command from a package instead of the entire package. For example, to load the single command **contourplot** from the **plots** package, type **with(plots, contourplot)**.

Suppressing Output. Certain commands produce output that might be considered superfluous. For example, when you assign a value to a variable *Maple* echoes the value. When you load a package using the **with** command, *Maple* prints a list of all the commands in the package. The output of a *Maple* command can always be suppressed by ending the command with a *colon* rather than a *semicolon*. Note that ending a statement with a colon has no effect on what *Maple* does internally; it only affects what you see in the Worksheet.

Problem Set A

Practice with *Maple*

In this problem set, you will use *Maple* to do some basic calculations, and then to plot, differentiate and integrate various functions. This problem set is the minimum you should do in order to reach a level of proficiency that will enable you to use *Maple* throughout the course. In this problem set you should concentrate on the mathematics, but if you wish to try out some of the editing features of the Worksheet interface, you can read about them in the next chapter.

1. Evaluate:

    (a) $3 + 9$

    (b) 2^{123}

    (c) π^2 and e to 35 digits

    (d) the fractions $\frac{22}{7}$, $\frac{311}{99}$, and $\frac{355}{113}$, and determine which is the best approximation to π.

2. Evaluate to ten digits:

    (a) $\dfrac{\sin(0.1)}{0.1}$

    (b) $\dfrac{\sin(0.01)}{0.01}$

    (c) $\dfrac{\sin(0.0001)}{0.0001}$.

3. Factor the polynomial $x^3 - y^3$.

4. Use the **plot** command to graph the following:

    (a) $y = 3x + 2$ for $-5 \le x \le 5$

    (b) $y = x^2 + x - 1$ for $-5 \le x \le 5$

    (c) $y = \sin x$ for $0 \le x \le 4\pi$

    (d) $y = \tan x$ for $-\pi/2 \le x \le \pi/2$

(e) $y = e^{-x^2}$ for $-2 \leq x \leq 2$.

5. Plot the functions x^4 and 2^x on the same graph, and determine how many times their graphs intersect. (*Hint*: You will probably have to make several plots, using intervals of various sizes, in order to find all the intersection points. You may have to limit the y-range in your plots.) Now find the values of the points of intersection by using the **fsolve** command.

6. The **contourplot** command takes a function of two variables (say $f(x, y)$) and plots several *level curves* of the function (that is, curves of the form $f(x, y) = c$, where c is a constant). For example, **contourplot(x^2 + y^2, x = −5..5, y = −5..5)** plots the level curves of the function $f(x, y) = x^2 + y^2$, in the square $-5 \leq x \leq 5$, $-5 \leq y \leq 5$. (Remember to enter **with(plots)** before using the **contourplot** command.)

(a) Use **contourplot** to plot level curves of the function $f(x, y) = 3y + y^3 - x^3$, in the region where x and y are between -1 and 1 (to get an idea of what the curves look like near the origin) and in a larger region (to get the big picture).

(b) A particular level set of a function can be plotted with the **implicitplot** command (also in the **plots** package). Plot the curve $3y + y^3 - x^3 = 5$.

(c) Plot the level set of the function $f(x, y) = y \ln x + x \ln y$ that contains the point $(1, 1)$.

7. The derivative of a function can be found by typing **diff(f(x), x)**. Find the derivatives of the following functions. In each case, if it appears that the answer *Maple* gives can be simplified, investigate the possibility by using **simplify**.

(a) $f(x) = 7x^3 + 3x^2 - 2x + 1$

(b) $f(x) = \dfrac{x + 1}{x^2 + 1}$

(c) $f(x) = \cos(x^2 + 1)$

(d) $f(x) = \arcsin(2x + 3)$

(e) $f(x) = \sqrt{1 + x^4}$

(f) $f(x) = x^r$

(g) $f(x) = \arctan(x)$.

8. Consider the polynomial $x^6 - 21x^5 + 175x^4 - 735x^3 + 1624x^2 - 1764x + 720$.

(a) Factor it.

(b) Find the roots using the *Maple* command **solve**.

(c) Plot the polynomial for $0.5 \leq x \leq 6.5$.

(d) Plot the polynomial and its derivative on the same interval. Explain your plot using calculus.

9. Use the *Maple* command **limit** to evaluate the following limits:

(a) $\displaystyle\lim_{x \to 0} \frac{\sin x}{x}$

(b) $\displaystyle\lim_{x \to \pi} \frac{1 + \cos x}{x - \pi}$

(c) $\displaystyle\lim_{x \to \infty} x\, e^{-x}$

(d) $\displaystyle\lim_{x \to 1^-} \frac{1}{x - 1}$

(e) $\displaystyle\lim_{x \to 0} \sin\left(\frac{1}{x}\right)$.

10. The integration command in *Maple* is **int**. Type **?int** to see the general syntax for both indefinite and definite integration. As an example, **int(sin(x), x)** gives the output $-\cos x$. Notice that the constant of integration is left out; *Maple* gives you one possible antiderivative of the function you specified. As an example of the definite integral, the command **int(sin(x), x = 0..Pi)** gives the answer 2.

See if *Maple* can do the following integrals. For the indefinite integrals, use *Maple* again to check the results by differentiating:

(a) $\int_0^{\pi/2} \sin x \, dx$

(b) $\int x \cos(x^2) \, dx$

(c) $\int \sin(3x)\sqrt{1 - \cos(3x)} \, dx$

(d) $\int \ln x \, dx$

(e) $\int x^2 \sqrt{x + 4} \, dx$

(f) $\int \sqrt{x^4 + 1} \, dx$

(g) $\int e^{\cos x} \, dx$

(h) $\int_{-\infty}^{\infty} e^{-x^2} \, dx$.

11. *Maple* can also integrate functions numerically, to get an approximate answer. This is useful in cases where no elementary formula exists for an antiderivative, or where *Maple* cannot perform the symbolic integration. The command that does this is **evalf(Int())**, which has the same syntax as **int**, but note the capital **I**. Find numerical values for the following integrals:

(a) $\int_0^\pi e^{\cos x}\,dx$

(b) $\int_0^1 \sqrt{x^4 + 1}\,dx$

(c) $\int_{-\infty}^\infty e^{-x^2}\,dx$.

For part (c), figure out how much the numerical answer differs from the exact answer found in the previous problem.

Chapter 4

Using *Maple* Worksheets

In this chapter we describe *Maple*'s Worksheet interface, which can be used to create an attractive and polished document. We describe the Worksheet interface for *Maple* V Release 4 running on *Windows 95*. Other platforms will be slightly different. Earlier releases of *Maple* are substantially different.

We suggest that you read this chapter while sitting at a computer with *Maple* running, trying out the various features as you read about them. You can also refer to the Sample Solutions at the end of this book to see examples of formatted *Maple* Worksheets.

Along the top of the *Maple* window is the *Menu Bar*, consisting of a row of *menu buttons* (**File**, **Edit**, *etc.*) We refer to these buttons using the convention that a button in a submenu of another menu is indicated by the name of the button in the top menu, then a colon, then the name of the button in the submenu, and so on. For example, **Insert:Execution Group:After Cursor** means the **After Cursor** button of the **Execution Group** submenu of the **Insert** menu. To get to this button, click on the **Insert** button at the top of the window, move the mouse pointer down to the **Execution Group** item, then move right to the submenu that appears, and select (click on) the **After Cursor** button.

Underneath the row of menu buttons is the *Tool Bar*, consisting of a row of buttons, most of which are shortcuts to menu commands. When **Help:Balloon Help** is turned on, descriptive text appears on the screen when you position the pointer over a Tool Bar button.

Below the Tool Bar is the *Context Bar*. The Context Bar contains buttons specific to certain parts of the *Maple* Worksheet. Its contents change according to the location of the cursor.

Organization of the Worksheet. A *Maple* Worksheet contains various types of information, including text, input, output and graphics. This information is organized into *Execution Groups*, *Paragraphs*, *Sections* and *Subsections*. Each of these is indicated by an extensible bracket along the the left edge of the Worksheet. Execution Groups are primarily for *Maple* input and the associated output and graphics. Paragraphs are for textual commentary as well as headers and titles. Sections and subsections are used to organize the Worksheet.

Execution Groups.   An Execution Group can contain input, output and graphics. The first line of an input region is indicated with a ">" prompt. On a color monitor, *Maple* input is red. Pressing the ENTER key while the cursor is in an input region of an Execution Group causes all of the input in the Execution Group to be evaluated; the cursor then jumps to the next Execution Group. The output generated by an Execution Group appears in the same Execution Group. Pressing SHIFT-ENTER in an input region produces a carriage return without causing evaluation.

A new Execution Group is inserted by clicking the [> button on the Tool Bar, or with **Insert:Execution Group:Before Cursor** or **Insert:Execution Group:After Cursor**. The new Execution Group contains a *Maple* prompt (>), indicating that you can type *Maple* input.

It is possible to insert text inside an Execution Group. We do not recommend it, since the effects can be unpredictable.

Inserting and Formatting Text.   A text region is called a Paragraph. Paragraphs can be used for commentary as well as for titles and headings. The most reliable way to insert a Paragraph is to click on [> in the Tool Bar to insert a new Execution Group, and then to click on **T** in the Tool Bar to change the Execution Group to a text region. You can also use **Insert:Paragraph:Before Cursor** or **Insert:Paragraph:After Cursor**, but these may have the undesirable effect of inserting a paragraph inside an Execution Group instead of creating a new region.

To change the format of an existing Paragraph, position the cursor in the Paragraph by clicking with the mouse. The Context Bar will contain three fields with drop lists, the first specifying a format for the Paragraph, the second a font style, and the third a font size. For example, the first field usually contains the word "Normal", specifying the Normal style of Paragraph. Clicking on the small arrow to the right of "Normal" causes a full list of styles to appear, and clicking on one of those styles causes the Paragraph to be reformatted. You can scroll up and down the list of styles by clicking on the arrows to the right of the list; there are predefined styles for title, author, lists, *etc*. The font style and size can be changed in a similar way. Further to the right are buttons that can be used to produce bold, italic or underlined text; these buttons can be applied to selected words or symbols in the text, as well as to the entire paragraph. Finally, the last three buttons in the Context Bar control the positioning of text: left-justified, centered or right-justified.

Sections and Subsections.   Execution Groups and Paragraphs can be grouped into sections or subsections. To make a section out of several contiguous Execution Groups and Paragraphs, select them by dragging the mouse across them (the selected regions will be darkened) and then select **Format:Indent**. A subsection is created by indenting regions inside a section. You can also create a new section or subsection by selecting the appropriate button from the **Insert** menu.

Sections and subsections are marked with a small button along the left edge of the Worksheet. Clicking on the button will "collapse" the section, making it invisible, and the symbol on the button will change from a minus sign to a plus sign. Clicking on the button again will make the section reappear.

Editing. The insertion point for text and input is marked by a blinking vertical bar. The insertion point is selected by positioning the pointer and clicking the mouse button, or by using the arrow keys on the keyboard to move the insertion point around.

The **Edit** menu contains basic editing commands, including **Cut**, **Copy**, and **Paste**. To cut or copy text, hold down the mouse button and drag the mouse across the relevant text, then select the appropriate button from the **Edit** menu. To paste material that you have cut or copied, move the pointer to the desired insertion point, click the mouse button (to set the insertion point), and choose **Paste** from the **Edit** menu.

To delete an entire region, position the cursor in the region. Then select **Edit:Delete Paragraph**. This works for deleting output and graphics as well as for text paragraphs.

Mathematical Typing. When inserting commentary in Worksheets, it is often useful to be able to type mathematical symbols, including integral signs, Greek letters, *etc*. *Maple* Worksheets allow you to insert formatted mathematical text using "inert" *Maple* commands. For example, to write the phrase "Consider the integral $\int_0^1 f(x)dx$" in Maple, do the following.

1. Create a text region with **Insert:Paragraph:After Cursor**.
2. Type the first part of the phrase, "Consider the integral".
3. Select **Insert:Maple Input**. A small box containing a question mark will appear in the text.
4. Type the *Maple* command associated with the mathematical expression you wish to display. In this case, the *Maple* command is **int(f(x), x = 0..1);**. As you type, the *Maple* expression will appear in the Context Bar, and the formatted expression will appear within the box in the text.
5. After you finish typing the mathematical formula, select **Insert:Text Input**, and then continue typing the text.
6. To edit the formula, select the formula and then edit it in the Context Bar.

Maple translates the names of Greek letters into symbols, so you can get Greek letters in text by selecting **Insert:Maple Input** and typing "delta", for example.

It is also possible to insert *active* Maple input in a text region, with **Insert:Math Input**, though we do not recommend it since the output still appears at the end of the Execution Group.

Manipulating Worksheets. To scroll through the Worksheet, use the "page up" and "page down" buttons on the keyboard, or drag the slider up and down in the scroll bar on the right side of the window.

You can have several Worksheets open at a given time. The **Window** menu contains a list of open Worksheets, and allows you to move among them. It also allows you to display two or more Worksheets simultaneously in various arrangements. Keep in mind that by default all the open Worksheets share the same underlying *Maple* process, so any variable assignments you make in one Worksheet will hold for the others as well.

Graphics.   When you click on a *Maple* graphic, a box appears around the graphic and the Menu Bar and Context Bar change, offering options and tools for manipulating the graphic. Here is a partial list of things you can do with a graphic.

1. Click in the graphic to display the coordinates of the pointer in a small box at the left side of the Context Bar.
2. Drag one of the small dots along the edge of the graphics box to resize the graphic.
3. Change the style of axes using the **Axes** menu.
4. Change the style of the graphic using the **Style** menu. If you change the style of a 3D graphic, you must double click on the graphic, or select the button labeled **R** from the Context Bar, to redraw the graphic.

Online Help.   In Chapter 3, we described how to access *Maple*'s online Help facility using the **?** prefix. The Help facility can also be accessed from the **Help** menu. For example, to see the documentation for a command you've typed in a Worksheet, select the command by clicking on it, and then choose **Help:Help On**. . .. The name of the command you selected appears in quotes in this menu button.

The *Maple* Help pages contain embedded cross-references called hyperlinks. Hyperlinks are underlined and displayed in green on a color monitor. Clicking on a hyperlink takes you to the indicated Help page.

Setting Defaults.   There are several ways to customize your *Maple* Worksheets, and once you become comfortable using *Maple* you may want to experiment with them.

If you are using a public computer and your Worksheet looks substantially different than you think it should, it may be because a previous user has changed some of the defaults. You may have to change the defaults back. If the computer you are using reboots from a central fileserver, you may be able to restore the defaults by rebooting the machine.

Memory Management.   *Maple* uses a lot of computer memory, for computations as well as for the interface. These memory demands can overwhelm some computers. Here are some ways to economize on memory.

1. Use the **restart** command. This command restarts the *Maple* kernel and clears all definitions.

2. Every open Worksheet uses memory, so Worksheets and Help pages should be closed when no longer needed.

3. Do each problem in a separate Worksheet. When you are finished with one problem, save and close the Worksheet, and then open up a new Worksheet for the next problem. You can print out each Worksheet separately, then bundle them together when you turn in your assignment. This strategy also limits the amount of work you lose if the Worksheet becomes corrupted.

4. If all else fails, save your Worksheet, quit *Maple*, and then restart it.

Save Your Work Often. *Maple* is a complex program, and as with any complex program, you will almost certainly experience unexpected problems and crashes while using it. The best protection against such problems is to save your work often. If *Maple* does crash, you'll only lose the work you've done since the last save.

Keyboard Equivalents. Some of the menu commands can be invoked from the keyboard. For example, the **Copy** command is invoked by holding down the CONTROL key and typing "C". These *keyboard equivalents* (called "Hot Keys" in the Maple on-line documentation) are efficient alternatives to using the mouse. You can read the keyboard equivalents off the menu buttons.

Here are two especially useful keyboard equivalents:

1. To insert a new Execution Group below the cursor, type CONTROL-J.

2. To insert a new Paragraph below the cursor, type CONTROL-J and then F5 (function key 5, along the top of the keyboard) or CONTROL-T.

Chapter 5

Solutions of Differential Equations

In this chapter, we show how to solve differential equations with *Maple*. For many differential equations, the *Maple* command **dsolve** produces the general solution to the differential equation, or the specific solution to an associated initial value problem. We also discuss the existence, uniqueness, and stability of solutions of differential equations. These are fundamental issues in the theory and application of differential equations. An understanding of them helps in interpreting and using results produced by *Maple*.

Finding Symbolic Solutions

Consider the differential equation

$$\frac{dy}{dx} = f(x, y). \tag{1}$$

A solution to this equation is a function $y(x)$ of the independent variable x that satisfies $y'(x) = f(x, y(x))$. It is sometimes possible to find a formula for the solutions to (1); we call such a formula a *symbolic solution*. In *Maple*, the command that finds symbolic solutions is **dsolve**. To find a symbolic solution of the differential equation (1), type

dsolve(diff(y(x), x) = f(x, y(x)), y(x));

Notice that you must type **y(x)** instead of **y** to let *Maple* know that y is the dependent variable and x is the independent variable. *Maple* produces the answer in terms of an arbitrary constant _C1. (For higher order equations, there will be several arbitrary constants.) For example, consider the differential equation

$$\frac{dy}{dx} = x + y.$$

You can solve this equation in *Maple* by typing

dsolve(diff(y(x), x) = x + y(x), y(x));

The output from this command is

$$y(x) = -x - 1 + e^x \, \_C1$$

You can then obtain specific solutions by choosing specific values for _C1. For example, the solution satisfying a given initial condition can be obtained by choosing _C1 appropriately. This value of _C1 can be found by imposing the initial condition on the general solution and solving for _C1. Alternatively, you can specify the initial condition as well as the differential equation when you use **dsolve**. To solve the initial value problem

$$\frac{dy}{dx} = f(x, y), \qquad y(x_0) = y_0,$$

you would type

> **dsolve({diff(y(x), x) = f(x, y(x)), y(x0) = y0}, y(x));**

For example, the command

> **sol1 := dsolve({diff(y(x), x) = x + y(x), y(0) = 2}, y(x));**

produces the output

$$sol1 := y(x) = -x - 1 + 3e^x$$

The solution of the differential equation is the expression following the equals sign. In this example, we have given the name **sol1** to the output.

Next, suppose you want to plot the solution, or find its value at a particular value of x. You can't just tell *Maple* to plot **y(x)** or evaluate **y(x)** at $x = 5$, say, because **dsolve** presents the solution in the form of an *equation*, not a function. The most straightforward way to plot or evaluate the solution is to define a function equal to the expression following the equals sign,

> **y1 := x -> -x - 1 + 3*exp(x);**

and then plot or evaluate **y1(x)**. You can also use the *Maple* commands **rhs** and **unapply** to extract the function automatically from **sol1**. The command

> **y1 := unapply(rhs(sol1), x);**

will extract the right-hand side (using **rhs**) of the solution equation and convert it (using **unapply**) into a function of the variable x.

We often want to study a family of solutions obtained by varying the initial condition. Here is a natural way to do this in *Maple*. Begin by solving the differential equation with a generic initial value. For example,

> **sol2 := dsolve({diff(y(x), x) = x + y(x), y(0) = c}, y(x));**

produces a solution formula in terms of the initial value c. The result of this command is

$$sol2 := y(x) = -x - 1 + e^x(1 + c)$$

Next, view the right-hand side as a function of both x and c. To define such a function in *Maple*, type

> **y2 := unapply(rhs(sol2), x, c);**

Now suppose we want to plot a family of solution curves with initial values $y(0) = -2, -1, \ldots, 4$ on the interval $0 \le x \le 3$. We can type

plot({y2(x, c) \$ c = −2..4}, x = 0..3);

The result of this command is shown in Figure 1. (The *Maple* syntax used here is the *sequence operator*, represented by a dollar sign. You can get more information about this operator by typing **?dollar** in a *Maple* session, or by consulting Chapter 3 and Chapter 8.)

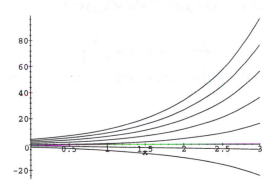

Figure 1

The **dsolve** command is robust; it can solve many differential equations. In fact, it can solve most of the differential equations that can be solved with the standard solution methods one learns in an introductory course. However, there are many other equations, some of which can be solved by more advanced solution methods, that *Maple* cannot solve.

REMARK. Although we have focused on first order equations in this section, **dsolve** also solves higher order equations and systems of equations. For examples, see Chapter 8, Chapter 12 and the Sample Worksheet Solutions.

Existence and Uniqueness

The fundamental existence and uniqueness theorems for differential equations guarantee that every initial condition $y(x_0) = y_0$ leads to a unique solution near x_0, provided that the right-hand side of the differential equation (1) is a "nice" function (to be specific, provided that f and $\partial f/\partial y$ are continuous functions). Graphically, these theorems say that there is a solution curve through every point, and that the solution curves cannot cross. Thus, initial value problems have exactly one solution, but, since there are an infinite number of initial conditions, differential equations have an infinite number of solutions. This principle is implicit in the results obtained above with **dsolve**; when we do not specify an initial condition, the solution depends on an arbitrary constant; when we specify an initial condition, the solution is completely determined.

It is important to remember that the existence and uniqueness theorems only guarantee the existence of a solution *near* the initial point x_0. Consider the initial value problem

$$\frac{dy}{dx} = y^2, \qquad y(0) = 1.$$

From the *Maple* command

sol3 := dsolve({diff(y(x), x) = y(x)^2, y(0) = 1}, y(x));

we get the output

$$sol3 := y(x) = -\frac{1}{x-1}$$

To understand where the solution exists, we can graph the expression given by **dsolve** in Figure 2.

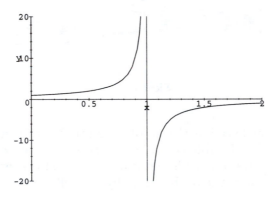

Figure 2

We see that the graph has two branches. Since the left branch passes through the initial data point $(0, 1)$, it is the desired solution. Furthermore, we see that the solution exists from $-\infty$ to 1, but does not extend beyond 1. In fact, it becomes unbounded as x approaches 1 from the left. Note that *Maple* includes the asymptote in its graph.

Stability of Differential Equations

In addition to existence and uniqueness, the sensitivity of the solution of an initial value problem to the initial value is a fundamental issue in the theory and application of differential equations. We introduce this issue through the following examples.

EXAMPLE 1. Consider the initial value problem

$$\frac{dy}{dx} + 2y = e^{-x}, \qquad y(0) = y_0.$$

We ask: How does the solution $y(x)$ depend on the initial value y_0? More specifically: Does the solution depend continuously on y_0? Do small variations in y_0 lead to small, or large, variations in the solution? Since the solution is

$$y(x) = e^{-x} + (y_0 - 1)e^{-2x},$$

we immediately see that, for any fixed x, the solution $y(x)$ depends continuously on y_0. We can say more. If we let $\tilde{y}(x)$ be the solution of the same equation, but with the initial condition $\tilde{y}(0) = \tilde{y}_0$, then $\tilde{y}(x) = e^{-x} + (\tilde{y}_0 - 1)e^{-2x}$. So, we have

$$|y(x) - \tilde{y}(x)| = |y_0 - \tilde{y}_0|e^{-2x}.$$

Thus for $x \geq 0$ we see that $|y(x) - \tilde{y}(x)|$ is never larger than $|y_0 - \tilde{y}_0|$. In fact, $|y(x) - \tilde{y}(x)|$ decreases as x increases. For $x \leq 0$ the situation is different. When x is a large negative number, the initial difference $|y_0 - \tilde{y}_0|$ is magnified by the large factor e^{-2x}. Even though $y(x)$ depends continuously on y_0, small changes in y_0 lead to large changes in $y(x)$. For example, if $|y_0 - \tilde{y}_0| = 10^{-3}$, then $|y(-7) - \tilde{y}(-7)| = 10^{-3}e^{14} \approx 1203$. These observations are confirmed by Figure 3, a plot of the solutions corresponding to initial values $y(0) = 0.97, 1, 1.03$ for $-3 \leq x \leq 3$.

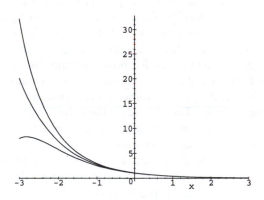

Figure 3

EXAMPLE 2. The solution of the initial value problem

$$\frac{dy}{dx} - 2y = -3e^{-x}, \qquad y(0) = y_0$$

is $y(x) = e^{-x} + (y_0 - 1)e^{2x}$. Again, $y(x)$ depends continuously on y_0 for fixed x. Letting $\tilde{y}(x)$ be the solution with initial value $\tilde{y}_0$, we see that

$$|y(x) - \tilde{y}(x)| = |y_0 - \tilde{y}_0|e^{2x}.$$

Now we see that $y(x)$ is very sensitive to changes in the initial value for x large and positive, but insensitive for x negative. These observations are confirmed by Figure 4, a plot of the solutions corresponding to initial values $y(0) = 0.97, 1, 1.03$ for $-2 \leq x \leq 2$.

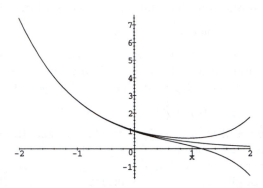

Figure 4

Often, we are primarily interested in positive values of x, such as when x corresponds to time in a physical problem. If an initial value problem is to predict the future of a physical system effectively, the solution for positive x should be fairly insensitive to the initial value; *i.e.*, small changes in the initial value should lead to small changes in the solution for positive time. To see the importance of this principle, note that for a physical system the initial value y_0 is typically not known exactly, but is found by measurement. When measuring y_0, we usually get only an approximate value $\tilde{y}_0$. Then, if $\tilde{y}(x)$ is the solution corresponding to $\tilde{y}_0$ and the solution is very sensitive to the initial value, $\tilde{y}(x)$ will have little relation to the actual state $y(x)$ of the system as x increases.

A differential equation (like the equation in Example 1) whose solution is fairly insensitive to the initial value as x increases is called *stable*, while a differential equation (like the equation in Example 2) whose solution is very sensitive to the initial value as x increases is called *unstable*. As we have seen, if an equation is unstable and is used over a long time interval, a small error in the initial value can result in a large error later on. Caution should be exercised when using an unstable equation to model a physical problem.

REMARK. We have considered stability as x increases; *i.e.*, stability to the right. We can also consider stability to the left. There are equations that are stable both to the left and to the right, and equations that are unstable both to the left and right.

The following theorem, which we state without proof, is often useful in assessing stability.

THEOREM. *Suppose $f(x, y)$ has continuous first order partial derivatives in the vertical strip*

$$R = \{(x, y) \ : \ x_0 \le x \le x_1, -\infty < y < \infty\},$$

and suppose there are numbers K and L such that

$$K \le \frac{\partial f}{\partial y}(x, y) \le L, \quad \text{for all } (x, y) \in R.$$

If $y(x)$ and $\tilde{y}(x)$ are solutions of $dy/dx = f(x, y)$ on the interval $x_0 \le x \le x_1$ with initial values $y(x_0) = y_0$ and $\tilde{y}(x_0) = \tilde{y}_0$, respectively, then

$$|y_0 - \tilde{y}_0|e^{K(x-x_0)} \le |y(x) - \tilde{y}(x)| \le |y_0 - \tilde{y}_0|e^{L(x-x_0)},$$

for all $x_0 \le x \le x_1$.

If $L \le 0$, then the right-hand inequality in the theorem shows that

$$|y(x) - \tilde{y}(x)| \le |y_0 - \tilde{y}_0|, \quad \text{for all } x_0 \le x \le x_1.$$

Thus the solutions differ by no more than the difference in the initial values, and the differential equation is stable. Moreover, if $L > 0$, but not too large, and $x_1 - x_0$ is not too large, then

$$|y(x) - \tilde{y}(x)| \le M|y_0 - \tilde{y}_0|, \quad \text{for all } x_0 \le x \le x_1,$$

where $M = e^{L(x_1-x_0)}$ is a moderate sized constant. Thus the equation is only mildly sensitive to changes in the initial value, and the equation is only mildly unstable. On the other hand, if $K > 0$, then the left-hand inequality in the theorem shows that the solution is sensitive to changes in the initial value, especially over long intervals. We can briefly summarize these results by saying that if $\partial f/\partial y \le 0$, then the differential equation is stable; but if $\partial f/\partial y > 0$, then the equation is unstable.

REMARK. The right-hand inequality in the theorem is an example of a *continuous dependence* result; it shows that the solution depends continuously on the initial value.

Let us examine our examples in light of these observations. Rewriting the equation in Example 1 as $dy/dx = -2y + e^{-x}$, we see that $f(x, y) = -2y + e^{-x}$ and $\partial f/\partial y = -2$. We can apply the theorem with $x_0 = 0, x_1 = \infty$, and $L = K = -2$ to find that

$$|y(x) - \tilde{y}(x)| = |y_0 - \tilde{y}_0|e^{-2x}, \quad \text{for all } x \ge 0,$$

as we found above from the solution formula. Thus the equation is stable. We could also conclude that the equation is stable just by noting that $\partial f/\partial y < 0$. Likewise, for the equation of Example 2, $f(x, y) = 2y - 3e^{-x}$ and $\partial f/\partial y = 2 > 0$, so the equation is unstable.

Of course, we can understand the stability of the differential equations in these two examples by examining the solution formulas. But $\partial f/\partial y$ can be calculated and its sign and size found, even if a solution formula cannot be found. For example, we can immediately tell that the differential equation $dy/dx + x^2 y^3 = \cos x$ is stable because

$$f(x, y) = -x^2 y^3 + \cos x$$

and $\partial f/\partial y = -3x^2 y^2 \leq 0$. Yet neither **dsolve** nor any other standard technique enables us to find a formula solution to this differential equation.

Throughout the course we will see many examples that illustrate the dependence (either sensitive or insensitive) of the solution to an initial value problem on the initial value.

Caveat.   It is important to realize that *Maple*, like any software system, may occasionally produce misleading or incorrect results. Consider, for example, the initial value problem

$$\frac{dy}{dx} = -xy^4, \qquad y(0) = 1.$$

In Version V.4 of *Maple*, the command

sol4 := dsolve({diff(y(x), x) = −x*y(x)^4, y(0) = 1}, y(x));

gives the output

sol4 :=

$$y(x) = \frac{2^{1/3}((3x^2 + 2)^2)^{1/3}}{3x^2 + 2}, \ -\frac{1}{2}\frac{2^{1/3}((3x^2 + 2)^2)^{1/3}}{3x^2 + 2} - \frac{1}{2}\frac{I\sqrt{3}\,2^{1/3}((3x^2 + 2)^2)^{1/3}}{3x^2 + 2}$$

suggesting that there are two solutions, in violation of uniqueness. However, it is easily seen that only the first function satisfies the initial condition. (Both functions satisfy the differential equation. The second function is complex-valued, and does not satisfy the initial condition.)   This example shows that a good theoretical understanding is a valuable guide in interpreting computer-generated results, and therefore in identifying situations in which *Maple* has produced a misleading or incorrect result.

Chapter 6

A Qualitative Approach to Differential Equations

In this chapter, we discuss a qualitative approach to the study of differential equations. With this approach, we obtain qualitative information about the solutions directly from the differential equation, without the use of a solution formula.

Consider the general first order differential equation

$$\frac{dx}{dt} = f(t, x).$$ (1)

We can obtain qualitative information about the solutions $x(t)$ by viewing (1) geometrically. Specifically, we can obtain this information from the direction field of (1). Recall that the direction field is obtained by drawing through each point (t, x) in the (t, x)-plane a short line segment with slope $f(t, x)$. Solutions, or integral curves, of (1) have the property that at each of their points they are tangent to the direction field at that point, and therefore the general *qualitative* nature of the solutions can be determined from the direction field. Direction fields can be drawn by hand for some simple differential equations, but *Maple* can draw them for any first order equation. We illustrate the qualitative approach with two examples.

EXAMPLE 1. Consider the equation

$$\frac{dx}{dt} = e^{-t} - 2x.$$ (2)

Maple's command for plotting direction fields is **dfieldplot**, in the **DEtools** package. To load this package, type **with(DEtools)**. You can then plot the direction field of (2) on the rectangle $-2 \leq t \leq 3, -1 \leq x \leq 2$ by typing

> **dfieldplot(diff(x(t), t) = exp(−t) − 2*x(t), x(t), t = −2..3, x = −2..3,**
> **arrows = LINE, axes = BOXED);**

The options are explained in the *Maple* online help. The resulting direction field is shown in Figure 1.

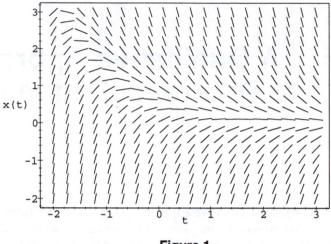

Figure 1

The direction field strongly suggests that if a solution is negative at some point, then it is increasing at that point, and that all solutions approach zero as $t \to \infty$. The general solution of equation (2) is $e^{-t} + ce^{-2t}$. Figure 2 shows the direction field together with several of these solutions.

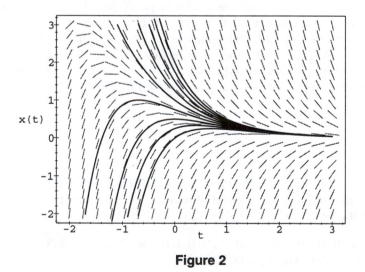

Figure 2

We can also deduce from the differential equation itself that the maximum points on the solution curves depicted in Figure 2 will occur along the curve $2x = e^{-t}$, since that is where $x' = 0$.

EXERCISE. Pursue this idea further by differentiating (2) and showing that the inflection points on the solution curves lie along the curve $4x = 3e^{-t}$.

Equation (2) can be solved explicitly because it is linear. Now let us consider an example that cannot be solved explicitly by any of the standard solution methods.

EXAMPLE 2. Consider the equation

$$\frac{dx}{dt} = x^2 + t. \tag{3}$$

Its direction field is obtained with the *Maple* command

dfieldplot(diff(x(t), t) = x(t)^2 + t, x(t), t = −2..2, x = −2..2,
    arrows = LINE, axes = BOXED);

and is shown in Figure 3.

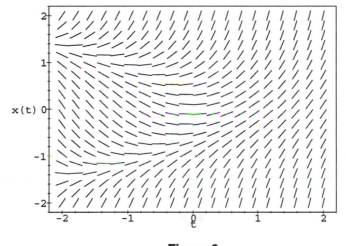

Figure 3

From the direction field, it appears that all solutions are increasing for $t > 0$, and that all solutions eventually become positive. Figure 3 also suggests that all solutions approach infinity. Does this happen at a finite value of t or only as $t \to \infty$? It is not easy to decide on the basis of Figure 3, but in fact for each solution $x(t)$ there is a finite value t^* such that $\lim_{t \to t^*} x(t) = \infty$. Later, we will again encounter equations that cannot be solved explicitly by formula techniques. We will apply a combination of qualitative and numerical tools to analyze such equations, and in particular establish the fact just asserted about the solutions to equation (3) approaching infinity in finite time.

EXERCISE. Use the *Maple* command **dfieldplot** to duplicate some of the direction fields in your textbook.

Autonomous Equations

Equations of the form

$$\frac{dx}{dt} = f(x), \tag{4}$$

which do not involve t in the right-hand side, are called *autonomous* equations. Autonomous equations represent physical systems whose rules of evolution do not change with time, and are particularly susceptible to qualitative analysis.

Consider equation (4); to be concrete, suppose $f(x)$ has two zeros x_1 and x_2, called *critical points* of the differential equation. Furthermore, suppose the graph of $f(x)$ is as shown in Figure 4.

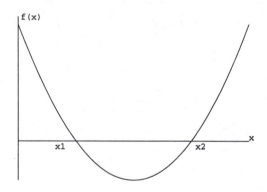

Figure 4

Then by considering the properties of $f(x)$, the direction field of (4) can be easily understood and drawn by hand. At the points $(0, x_1)$ and $(0, x_2)$ the slope is zero; at points $(0, x)$ with $x < x_1$, the slope is positive and increases from 0 to ∞ as x decreases from x_1 to $-\infty$; at points $(0, x)$ with $x_1 < x < x_2$, the slope is negative, and first decreases and then increases as x increases from x_1 to x_2; at points $(0, x)$ with $x > x_2$ the slope is positive and increases from 0 to ∞ as x increases from x_2 to ∞; finally, the slopes along any horizontal line are constant (since f does not depend on t). The direction field thus has the general appearance shown in Figure 5. Note that Figure 4 depicts $f(x)$ vs. x, while Figure 5 shows x vs. t.

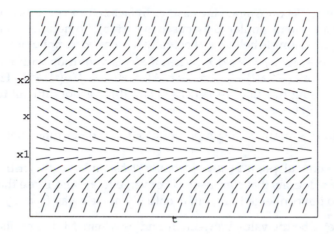

Figure 5

Several facts are suggested by the direction field, namely, that: (i) there are two constant solutions $x = x_1$ and $x = x_2$; (ii) solutions starting above x_2 tend to ∞; and (iii) all other solutions tend to x_1 as $t \to \infty$. In fact, we can derive these properties of the solutions $x(t)$ directly from the graph of $f(x)$ as follows.

(a) The constant functions $x(t) = x_1$ and $x(t) = x_2$ are solutions, called the *equilibrium solutions*. They are the unique solutions that satisfy the initial conditions $x(0) = x_1$ and $x(0) = x_2$, respectively.

(b) Consider a solution $x(t)$ with $x(0) < x_1$. Then, because of the uniqueness of solutions, it must be that $x(t) < x_1$ for all t and

$$x'(t) = f(x(t)) > 0.$$

Hence $x(t)$ is an increasing function. Since it is also bounded, it has a limit at infinity

$$\lim_{t \to \infty} x(t) = b \le x_1. \tag{5}$$

Could it happen that $b < x_1$? The answer is no, because if $b < x_1$, then

$$x'(t) = f(x(t)) \to f(b) > 0. \tag{6}$$

But equations (5) and (6) say that $x(t)$ approaches a horizontal asymptote at the same time that its slope is "permanently" bigger than a positive number. The resulting contradiction guarantees that

$$\lim_{t \to \infty} x(t) = x_1.$$

(c) If $x(t)$ is a solution with $x(0) > x_2$, then $x(t) > x_2$ for all t. Also

$$x'(t) = f(x(t)) > 0.$$

Hence $x(t)$ is once again increasing. In fact $x(t) \to \infty$. This can be shown by an argument similar to that used in part (b). Sometimes, as we shall see from the example below, $x(t)$ actually reaches ∞ in finite time.

(d) Now consider $x_1 < x_0 < x_2$. Then if $x(t)$ is a solution with $x(0) = x_0$, we must have $x_1 < x(t) < x_2$ for all t. Also, $x'(t) = f(x(t)) < 0$. Hence $x(t)$ is a decreasing function. Exactly as in part (b), it is not difficult to show that $\lim_{t \to \infty} x(t) = x_1$.

Note that in this analysis the properties of $x(t)$ are determined solely from the sign of $f(x)$.

We call the critical point x_1 *asymptotically stable* and call the critical point x_2 *unstable*, since solutions that start near x_1 converge to x_1, and those that start near x_2 not only do not converge to x_2, but eventually move away from x_2.

EXERCISE. Let $\bar{x}$ be the value between x_1 and x_2 where f takes on its minimum. Show that if $\bar{x} < x(0) < x_2$, then $x(t)$ has an inflection point where $x(t) = \bar{x}$.

Now we examine a specific example.

EXAMPLE 3. Consider the equation

$$\frac{dx}{dt} = x^2 - x.$$

Its direction field has the general appearance of Figure 5, with $x_1 = 0$ and $x_2 = 1$. But we can actually derive an explicit formula solution. If $x \neq 0$ and $x \neq 1$, we can solve by separating variables:

$$t + C = \int dt = \int \frac{dx}{x^2 - x}$$

$$= \int \left(\frac{1}{x - 1} - \frac{1}{x} \right) dx$$

$$= \ln|x - 1| - \ln|x|$$

$$= \ln \left| \frac{x - 1}{x} \right|$$

$$= \ln \left| 1 - \frac{1}{x} \right|.$$

Exponentiating both sides, we obtain

$$1 - \frac{1}{x} = ke^t,$$

where $k = \pm e^C$. Solving for x, we find that $x(t) = 1/(1 - ke^t)$. Noting that $x_0 = x(0) = 1/(1 - k)$ we have

$$x(t) = \frac{x_0}{(1 - x_0)e^t + x_0}. \tag{7}$$

Although formula (7) was derived under the assumption that $x_0 \neq 0, 1$, it is easily seen to be valid for all values of x_0.

Figure 6 contains the actual solution curves for the differential equation—as drawn by *Maple*.

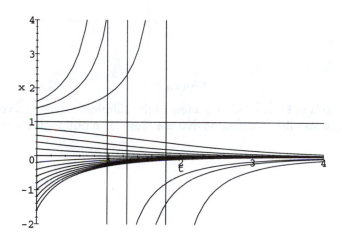

Figure 6

EXERCISES.

(a) Use formula (7) to verify the properties of $x(t)$ obtained above by the qualitative method. Determine where the solutions are increasing or decreasing, and whether there are limits as $t \to \infty$.

(b) Suppose $x_0 > 1$. Use formula (7) to show that $x(t) \to \infty$ in finite time. Find the time t^* at which this happens.

(c) Show that $x(t) = 0$ and $x(t) = 1$ are the equilibrium solutions. Does formula (7) yield these solutions?

Here is another example. The left graph in Figure 7 shows a function $f(x)$ with three zeros; thus, the autonomous differential equation $x' = f(x)$ has three critical points. Let's look at the critical point in the middle. Solutions starting below b have a positive slope, so they increase toward b. Solutions starting above b have a negative slope, so they decrease toward b. Therefore, we expect b to be an asymptotically stable equilibrium solution. A similar analysis of the signs of $f(x)$ suggests that the other two critical points are unstable.

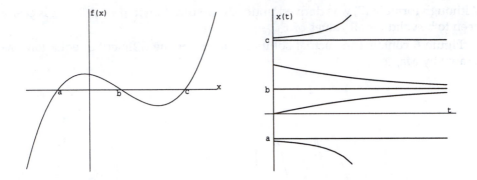

Figure 7

The right graph in Figure 7 shows a few of the solution curves. As you can see, this graph illustrates the conclusions that we drew from our analysis of the graph of $f(x)$.

Problem Set B

First Order Equations

1. The initial value problem

$$xy' + 2y = \sin x, \qquad y(\pi/2) = 1$$

has the solution

$$y(x) = x^{-2} \left(\frac{\pi^2}{4} - 1 - x \cos x + \sin x \right).$$

Define this function in *Maple* and then:

(a) Make a list of the values of $y(x)$ at $x = 0.5, 1, 1.5, 2, \ldots, 5$. (This can be done with the sequence operator. For more information, type **?dollar** in a *Maple* session. Also, recall the use of **evalf** to evaluate algebraic expressions numerically.)

(b) Graph $y(x)$ on the intervals $0 < x \le 2$, $1 \le x \le 10$, and $10 \le x \le 100$. Describe the behavior of the solution near $x = 0$ and for large values of x.

(c) Find the solutions $y_j(x)$ of the differential equation corresponding to the initial conditions

$$y_j(\pi/2) = 0.2j, \qquad j = 1, \ldots, 5.$$

Plot the functions $y_j(x)$, $j = 1, \ldots, 5$, on the same graph.

(d) What do the solutions have in common near $x = 0$? for large values of x? Can you find a solution to the differential equation that has no singularity at $x = 0$? If so, graph it.

2. Consider the initial value problem

$$xy' + y = 2x, \qquad y(1) = c.$$

(a) Solve it using *Maple*.

(b) Use *Maple* to graph the solutions for $c = 0.8, 0.9, 1, 1.1, 1.2$ on the interval $(0.75, 1.25)$.

(c) Evaluate these solutions at $x = 0.01, 0.1, 1, 10$.

(d) Now plot all five solutions together on the interval $(0, 2.5)$. How do changes in the initial data affect the solution as $x \to \infty$? as $x \to 0^+$?

(The sample solution for Problem 1 in the back of the book shows how to construct a sequence of values with nonintegral increments. See also the section *Sequences, Sets, and Lists* in Chapter 3.)

3. Solve the initial value problem

$$y' - y = \cos x, \qquad y(0) = c.$$

Use *Maple* to graph solutions for $c = -0.9, -0.8, \ldots, -0.1, 0$. Display all the solutions on the same interval between $x = 0$ and an appropriately chosen right endpoint. Explain what happens to the solution curves for large values of x. (*Hint*: You should identify three distinct types of behavior.) Now, based on this problem, and the material in Chapters 5 and 6, discuss what effect small changes in initial data can have on the global behavior of solution curves. (See the note at the end of Problem 2.)

4. Consider the differential equation

$$\frac{dy}{dx} = \frac{x - e^{-x}}{y + e^y}$$

(*cf.* Problem 7, Section 2.3 in Boyce & DiPrima).

(a) Solve it using *Maple*. Observe that the solution is given implicitly in the form

$$f(x, y) = c.$$

(b) Use **contourplot** (see the online help) to see what the solution curves look like. For your x and y ranges, you might use x = −1..3 and y = −2..2.

(c) Use **implicitplot** to plot the solution satisfying the initial condition $y(1.5) = 0.5$. Your plot should show two curves. Indicate which one corresponds to the solution.

(d) Given a value x_1, to find $y(x_1)$ you need to find the solution of

$$f(x_1, y) = f(1.5, 0.5)$$

(viewed as an equation in y) near $y = 0.5$. This can be done with the **fsolve** command. Find $y(0), y(1), y(1.8), y(2.1)$. Mark these values on your plot.

5. Consider the differential equation

$$(e^x \sin y - 2y \sin x)\, dx + (e^x \cos y + 2 \cos x)\, dy = 0,$$

(*cf.* Problem 7, Section 2.8 in Boyce & DiPrima).

(a) Solve it using *Maple*. Observe that the solution is given implicitly in the form

$$f(x, y) = c.$$

(b) Use **contourplot** to see what the solution curves look like. For your x and y ranges, you might use x = −3..3 and y = −3..3.

(c) Use **implicitplot** to plot the solution satisfying the initial condition $y(0) = 0.5$. Your plot may show several curves. Indicate which one corresponds to the solution.

(d) Given a value x_1, to find $y(x_1)$ you need to find the solution of

$$f(x_1, y) = f(0, 0.5)$$

(viewed as an equation in y) near $y = 0.5$. This can be done with the **fsolve** command. Find $y(-1), y(1), y(2)$. Mark these values on your plot.

6. In this problem, we study continuous dependence of solutions on initial data.

   (a) Solve the initial value problem

   $$y' = y/(1 + x^2), \qquad y(0) = c.$$

   (b) Let y_c denote the solution in part (a). Use *Maple* to plot the solutions y_c for $c = -10, -9, \ldots, -1, 0, 1, \ldots, 10$ on one graph. Display all the solutions on the interval $-20 \le x \le 20$.

   (c) Compute $\lim_{x \to \pm\infty} y_1(x)$.

   (d) Now find a single constant L such that for all real x, we have

   $$|y_a(x) - y_b(x)| \le L|a - b|,$$

   for any pair of numbers a and b. Show that $L = 1$ will work if we consider only negative values of x.

7. Use **dsolve** to solve the following differential equations or initial value problems from Boyce & DiPrima. In some cases, *Maple* will not be able to solve the equation. (Before moving on to the next equation, make sure you haven't mistyped something.) In other cases, *Maple* may give extraneous solutions. (Sometimes these correspond to imaginary roots of the equation it solves to get y in terms of x.) If so, you should indicate which solution or solutions are valid. You also might try entering alternative forms of an equation, for example, $M + Ny' = 0$ instead of $y' = -M/N$, or vice versa.

   (a) $x^3 y' + 4x^2 y = e^{-x}, y(-1) = 0$ (Sect. 2.1, Prob. 19)

   (b) $y' = ry - ky^2$ (Sect. 2.2, Prob. 39)

   (c) $y' = x(x^2 + 1)/4y^3, y(0) = -1/\sqrt{2}$ (Sect. 2.3, Prob. 16)

   (d) $(e^x \sin y + 3y)\, dx - (3x - e^x \sin y)\, dy = 0$ (Sect. 2.8, Prob. 8)

(e) $dy/dx = (2y - x)/(2x - y)$ (Sect. 2.9, Prob. 15(a))

(f) $dy/dx = (2x + y)/(3 + 3y^2 - x), y(0) = 0$ (Sect. 2.10, Prob. 3).

8. Use **dsolve** to solve the following differential equations or initial value problems from Boyce & DiPrima. See Problem 7 for additional instructions.

(a) $xy' + (x + 1)y = x, y(\ln 2) = 1$ (Sect. 2.1, Prob. 20)

(b) $y' - (1/x)y = x^{1/2}$ (Sect. 2.2, Prob. 19)

(c) $y' = 2(1 + x)(1 + y^2), y(0) = 0$ (Sect. 2.3, Prob. 26)

(d) $(x \ln y + xy) dx + (y \ln x + xy) dy = 0; x > 0, y > 0$ (Sect. 2.8, Prob. 11)

(e) $(x^2 + 3xy + y^2) dx - x^2 dy = 0$ (Sect. 2.9, Prob. 8)

(f) $dy/dx = -(2xy + y^2 + 1)/(x^2 + 2xy)$ (Sect. 2.10, Prob. 5).

9. Chapter 6 describes how to plot the direction field for a first order equation. For each equation below, plot the direction field on a rectangle large enough (but not too large) to show clearly all of its equilibrium points. Find the equilibria and state whether each is stable or unstable. If you cannot determine the precise value of an equilibrium point from the direction field, use **fsolve**.

(a) $y' = -y(y - 2)(y - 4)/10$

(b) $y' = y^2 - 3y + 1$.

(c) $y' = 0.1y - \sin y$.

10. In this problem, we use the direction field capabilities of *Maple* to study two nonlinear equations, one autonomous and one nonautonomous.

(a) Plot the direction field for the equation

$$\frac{dy}{dx} = 3 \sin y + y - 2$$

on a rectangle large enough (but not too large) to show all possible limiting behaviors of solutions as $x \to \infty$. Find approximate values for all the equilibria of the system (you should be able to do this with **fsolve** using guesses based on the direction field picture), and state whether each is stable or unstable.

(b) Plot the direction field for the equation

$$\frac{dy}{dx} = y^2 - xy,$$

again using a rectangle large enough to show the possible limiting behaviors. Identify the unique constant solution. Why is this solution evident from the differential equation? If a solution curve is ever below the constant solution, what must its limiting behavior be as $x \to +\infty$? For solutions lying above the constant solution, describe two possible limiting behaviors as $x \to +\infty$.

There is a solution curve which lies along the boundary of the two limiting behaviors. Explain (from the differential equation) why that solution curve cannot be bounded. Can you guess how its limiting behavior might differ from that of the two limiting behaviors you have identified?

11. The solution of the differential equation

$$y' = \frac{2y - x}{2x - y}$$

is given implicitly by $|x - y| = c|x + y|^3$ (Boyce & DiPrima, Problem 15a, Section 2.9; you need not verify the solution with **dsolve**.) However, it is difficult to understand the solutions directly from this algebraic information.

(a) Use **dfieldplot** to plot the direction field in several different rectangles.

(b) Use **implicitplot** to plot the solutions with initial conditions $y(2) = 1$ and $y(0) = -3$. (Note that **abs** is the absolute value function in *Maple*.) Use the **display** command to put these plots and the vector field plot together on the same graph. (Two plots named **r** and **s** can be combined with the command **display({r, s})** after loading the *Maple* **plots** package. See Chapter 8 and the online help for additional information.)

(c) For the two different initial conditions in part (b), use your pictures to estimate the largest interval on which the unique solution function is defined.

12. Consider the differential equation

$$x' = -tx^2.$$

(a) Use *Maple* to plot the direction field of the differential equation. Is there a constant solution? If $x(0) > 0$, what happens as t increases? If $x(0) < 0$, what happens as t increases?

(b) Use **dsolve** to solve the differential equation, thereby obtaining a solution of the form

$$x(t) = \frac{2}{t^2 + 2c}.$$

Note that $x(0) = 1/c$. In graphing $x(t)$, it helps to consider three separate cases: (i) $c > 0$, (ii) $c = 0$, and (iii) $c < 0$. In each case, graph the solution for several specific values of c, and identify important features of the curves. In particular, in case (iii), compute the location of the asymptotes in terms of c.

(c) Identify five different types of solution curves for the differential equation. In each case, specify the t-interval of existence, whether the solution is increasing or decreasing, and the limiting (asymptotic) behavior at the ends of the interval.

(d) Use the **display** command to combine the direction field plotted in (a) with the plot of a solution in case (iii). (Two plots named **r** and **s** can be combined with the command **display({r, s})** after loading the *Maple* **plots** package. See Chapter 8 and the online help for additional information.)

13. Consider the following logistic-with-threshold model for population growth:

$$x' = x(1 - x)(x - 3).$$

(a) Find the equilibrium solutions of the differential equation. Now draw the direction field, and use it to decide which equilibrium solutions are stable and which are unstable.

(b) Next replace the logistic law by the Gompertz model, but retain the threshold feature. The equation becomes

$$x' = x(1 - \ln x)(x - 3).$$

Once again, find the equilibrium solutions and draw the direction field. You may have trouble near $x = 0$ (because of the logarithm). Use a range like $0.01 \leq x \leq 4$ to circumvent that problem. But you will also have difficulty "reading the field" between 2.5 and 3. There appears to be a continuum of equilibrium solutions.

(c) Plot the function $f(x) = x(1 - \ln x)(x - 3)$ on the interval $0.01 \leq x \leq 4$, and then use *Maple's* **limit** command to evaluate $\lim_{x \to 0} f(x)$.

(d) Use these plots and the discussion in Chapter 6 to decide which equilibrium solutions are stable and which are unstable. Now use the last plot to explain why the direction field (for $2.5 \leq x \leq 3$) appears so inconclusive regarding the stability of the equilibrium solutions.

14. This problem is based on Example 3 in Section 2.5 of Boyce & DiPrima: "A tank contains Q_0 lb of salt dissolved in 100 gal of water. Water containing $\frac{1}{4}$ lb of salt per gallon enters the tank at a rate of 3 gal/min, and the well-stirred solution leaves the tank at the same rate. Find an expression for the amount of salt $Q(t)$ in the tank at time t."

The differential equation

$$Q'(t) = 0.75 - 0.03Q(t)$$

models the problem (*cf.* equation (17) in Section 2.5).

(a) Plot the right-hand side of the differential equation as a function of Q, and identify the critical point.

(b) Analyze the long-term behavior of the solution curves by examining the sign of the right-hand side of the differential equation, in a similar fashion to the discussion in Chapter 6.

(c) Use *Maple* to plot the direction field of the differential equation. In choosing the rectangle for the direction field be sure to include the point $(0, 0)$ and the critical value of Q.

(d) Use **dsolve** to find the solution $Q(t)$ and plot it for several specific values of Q_0. Do the solutions behave as indicated in parts (b) and (d)? (You might try combining a direction field plot with that of the solution curves.)

15. A 10-gallon tank contains a mixture consisting of 1 gallon of water and an undetermined number $S(0)$ of pounds of salt in the solution. Water containing 1 lb/gal of salt begins flowing into the tank at the rate of 2 gal/min. The well-mixed solution flows out at a rate of 1 gal/min. Derive the differential equation for $S(t)$, the number of pounds of salt in the tank after t minutes, that models this physical situation. (Note: At time $t = 0$ there is 1 gallon of solution, but the volume increases with time.) Now draw the direction field of the differential equation on the rectangle $0 \le t \le 10$, $0 \le S \le 10$. From your plot,

(a) find the value A of $S(0)$ below which the amount of salt is a constantly increasing function; but above which, the amount of salt will temporarily decrease before increasing;

(b) indicate how the nature of the solution function in case $S(0) = 1$ differs from all other solutions.

Now use **dsolve** to solve the differential equation. Reinforce your conclusions above by

(c) algebraically computing the value of A;

(d) giving the formula for the solution function when $S(0) = 1$;

(e) giving the amount of salt in the tank (in terms of $S(0)$) when it is at the point of overflowing;

(f) computing, for $S(0) > A$, the minimum amount of salt in the tank, and the time it occurs;

(g) explaining what principle guarantees the truth of the following statement: If two solutions S_1, S_2 correspond to initial data $S_1(0)$, $S_2(0)$ with $S_1(0) < S_2(0)$, then for any $t \ge 0$, it must be that $S_1(t) < S_2(t)$.

16. In this problem, we use **dsolve** and either **solve** or **fsolve** to model some population data. The procedure will be:

    (i) Assume a model differential equation involving unknown parameters.

    (ii) Use **dsolve** to find the solution of the differential equation in terms of the parameters.

    (iii) Use **solve** or **fsolve** to find the values of the parameters that fit the given data.

    (iv) Make predictions based on the results of the previous steps.

(a) Let's use the model

$$\frac{dp}{dt} = ap + b, \quad p(0) = c,$$

where p represents the population at time t. Check to see that **dsolve** can solve this initial value problem in terms of the unknown constants a, b, c. Then define a function that expresses the solution at time t in terms of a, b, c, and t.

(b) Next, let's try to model the population of Nevada, currently the fastest growing state in the U.S. Here is a table of census data:

Year	Population in thousands
1950	160.1
1960	285.3
1970	488.7
1980	800.5
1990	1201.8

We would like to find the values of a, b, and c that fit the data. However, with three unknown constants we will not be able to fit five data points. Use **solve** or **fsolve** to find the values of a, b, and c that give the correct population for the years 1960, 1970, and 1980. We will later use the data from 1950 and 1990 to check the accuracy of the model. *Important:* In this part let t represent the time in years since 1950, because *Maple* may get stuck if you ask it to fit the data at such high values of t as 1960–1980. If you use **fsolve**, you will also need to specify ranges to guarantee that its algorithm converges. You might specify **a** = 0..1, for example. You could also figure out a reasonable range for c and specify that as well.

(c) Now define a function of t that expresses the predicted population in year t using the values of a, b, c found in part (b). Find the population this model gives for 1950 and 1990, and compare with the values in the table above. Use the model to predict the population of Nevada in the year 2000, and to predict when the population will reach 3 million. How would you adjust these predictions based on the 1950 and 1990 data? Finally, graph the population the model gives from 1950 to 2050, and describe the predicted future of the population of Nevada, including the limiting population (if any) as $t \to \infty$.

17. Consider $x' = (\alpha - 1)x - x^3$.

(a) Use **solve** to find the roots of $(\alpha - 1)x - x^3$. Explain why $x = 0$ is the only real root when $\alpha \le 1$, and why there are three distinct real roots when $\alpha > 1$.

(b) For $\alpha = -2, -1, 0$, draw a direction field for the differential equation and deduce that there is only one equilibrium solution. What is it? Is it stable?

(c) Do the same for $\alpha = 1$.

(d) For $\alpha = 1.5, 2$, draw the direction field. Identify all equilibrium solutions, and describe their stability.

(e) Explain the following statement: "As α increases through 1, the stable solution $x = 0$ *bifurcates* into two stable solutions."

Chapter 7

Numerical Methods

Even though many differential equations can be solved explicitly in terms of the elementary functions of calculus, many others cannot. Consider, for example, the equation

$$\frac{dy}{dx} = e^{-x^2}.$$

Its solutions are the integrals, or antiderivatives, of e^{-x^2},

$$y(x) = \int e^{-x^2}\, dx + C,$$

but it is known that these integrals cannot be expressed in terms of the elementary functions of calculus. Similarly, the solution to the associated initial value problem

$$\frac{dy}{dx} = e^{-x^2}, \qquad y(x_0) = y_0$$

can be written as

$$y(x) = y_0 + \int_{x_0}^{x} e^{-s^2}\, ds,$$

but this formula cannot be simplified further; specifically, it is not an elementary function. (An *elementary function* is defined to be a polynomial, power, trigonometric, inverse trigonometric, exponential, or logarithmic function, or any combination of them via the processes of algebra and of composition and inversion.) More dramatic is the "simple" equation

$$\frac{dy}{dx} = x + y^2.$$

The solutions to this equation cannot be written in terms of elementary functions, or even in terms of repeated integrals of elementary functions!

When we cannot find an antiderivative in terms of elementary functions, we turn to a numerical methods such as the trapezoid rule or Simpson's rule. Similarly, when we cannot solve a differential equation explicitly, or if the formula we find is too complicated, we also turn to numerical methods to solve the initial value problem.

Numerical Solutions Using *Maple*

Suppose we are interested in finding the solution to the initial value problem

$$\frac{dy}{dx} = f(x, y), \qquad y(x_0) = y_0,$$

and suppose that we have no formula for $y(x)$. Then we have no obvious way to calculate $y(x)$ for particular values of x. In such a situation, our strategy will be to produce a function $y_a(x)$ that both is a good approximation to $y(x)$ and can be calculated. The subscript a on y_a stands for *approximate solution*. Such a function y_a can be found with the *Maple* command **dsolve(. . ., numeric)**, *Maple*'s numerical differential equation solver. We illustrate the use of **dsolve(. . ., numeric)** by considering the initial value problem

$$\frac{dy}{dx} = \frac{x}{y}, \qquad y(0) = 1. \tag{1}$$

Its exact solution is easily found to be

$$y(x) = \sqrt{x^2 + 1}.$$

Since we have an explicit formula for this solution, we will be able to compare the approximate solution $y_a(x)$ and the exact solution $y(x)$. To obtain $y_a(x)$, type

   ivp := {diff(y(x), x) = x/y(x), y(0) = 1};
   sol := dsolve(ivp, y(x), numeric, maxfun = 1000, startinit = TRUE);

The first argument to **dsolve** is a set that specifies the differential equation and the initial condition. The second argument specifies the dependent variable y and the independent variable x. The third argument tells *Maple* to use the numerical solver instead of the symbolic one. The fourth argument, **maxfun = 1000**, is an option limiting the number of computations *Maple* will do to produce a numerical solution; this important option is explained in detail later. The final option will also be explained later, in the section *Further Remarks on dsolve(...,numeric), plot and DEplot*.

The preceding commands produce the output

       `sol := `**proc**`(rkf45_x)` . . . **end**

The output means that **sol** is a *Maple* procedure that computes an approximate solution to our initial value problem. The expression "rkf45" refers to the particular numerical method used by **dsolve(. . ., numeric)**. To get the value of the approximate solution at a particular point, say at $x = 1.5$, type **sol(1.5)**, which produces $[x = 1.5, y(x) = 1.802775648877327]$. Note that $y(1.5) = \sqrt{13}/2 = 1.802775637...$, so that the approximate value is correct to eight digits.

To extract a standard function that just returns the y-value, we type

   ya := u -> subs(sol(u), y(x));

Now **ya(1.5)** produces 1.802775648877327. We can make a table of values to compare the numerical solution with the actual solution. The commands

> onerow := x –> [x, evalf(ya(x)), evalf(sqrt(x^2 + 1))];
> rowlist := map(onerow, [0.2*j $ j = 1..10]):
> map(print, rowlist):

produce a table of numerical solution values and exact solution values at the points $0, 0.2, \ldots, 1.8, 2$. The result is shown in Table 1. (See Chapter 8 for more information about lists and the **map** command.)

x	$ya(x)$	$y(x)$
0.2	1.019803907	1.019803903
0.4	1.077032964	1.077032961
0.6	1.166190383	1.166190379
0.8	1.280624855	1.280624847
1.0	1.414213571	1.414213562
1.2	1.562049944	1.562049935
1.4	1.720465063	1.720465053
1.6	1.886796237	1.886796226
1.8	2.059126041	2.059126028
2.0	2.236067990	2.236067978

Table 1

We see that we have at least eight correct digits in all cases. So **ya(x)** is indeed a very good approximation to $y(x)$.

A graph of the approximate solution on the interval $[0, 2]$ can be obtained with the command

> plot(ya, 0..2);

The result is shown in Figure 1.

EXERCISE. Display both **ya(x)** and the exact solution $y(x) = \sqrt{x^2 + 1}$ on the same graph. Can you distinguish between the two curves?

We have illustrated **dsolve(..., numeric)** on the initial value problem (1), which can easily be solved explicitly. We can just as readily apply **dsolve(..., numeric)** to any first order initial value problem. Consider, for example,

$$\frac{dy}{dx} = x + y^2, \qquad y(0) = 1, \tag{2}$$

which cannot be solved in terms of elementary functions. The commands

> ivp2 := {diff(y(x), x) = x + y(x)^2, y(0) = 1};
> sol2 :=dsolve(ivp2, y(x), numeric, maxfun = 1000, startinit = TRUE);
> ya2 := u –> subs(sol2(u), y(x));

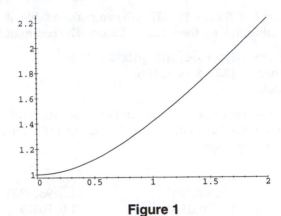

Figure 1

will produce an approximate solution.

EXERCISE. The solution $y(x)$ to (2) is increasing for x positive, and there is a finite value x^* such that $\lim_{x \to x^*} y(x) = \infty$; the interval of existence of the solution is $(-\infty, x^*)$. Graph the numerical solution. Does **plot** reveal the value of x^*?

When **dsolve(. . ., numeric)** computes a numerical solution at a point x, it marches from the initial point x_0 to x in a sequence of steps. If the exact solution is changing very rapidly, then **dsolve(. . ., numeric)** will take very small steps in order to stay close to the solution. This will be the case, for example, when the solution is oscillatory or singular (*i.e.*, has a vertical asymptote).

Each step involves several function evaluations. By default, the maximum number of function evaluations allowed is $30,000$. Such a large number of steps may take a long time, even on a computer; moreover, when there is an asymptote, following the solution for $30,000$ steps can lead to such large solution values that the numerical procedure fails. It is thus useful to limit the maximum number of steps allowed by **dsolve(. . ., numeric)** to a smaller number. This can be done with the **maxfun** option. For many problems, **maxfun = 1000** is an appropriate setting. On rare occasions, as when computing an oscillatory or singular solution, you may have to increase or decrease the value of **maxfun**.

EXERCISE. Use **dsolve(. . ., numeric)** with **maxfun = 1000** to calculate a numerical solution to

$$\frac{dy}{dx} = y\sin(x^2), \qquad y(0) = 1.$$

Now try to evaluate the numerical solution at $x = 5$ and $x = 10$. Can you explain why the numerical procedure generated by **dsolve(. . ., numeric)** fails at $x = 10$? Now compute a new numerical solution with **maxfun = 3000**, and evaluate this solution at $x = 10$. What happens? How far to the right can you go before **dsolve(. . ., numeric)** uses up its allotted 3000 steps?

REMARK.   Although we have focused on first order equations in this chapter, **dsolve(..., numeric)** also works on higher order equations and systems of equations. Examples can be found in Chapter 8, Chapter 12, and the Sample Worksheet Solutions.

Some Elementary Numerical Methods

In order to give some idea how **dsolve(..., numeric)** calculates approximate values, we now discuss three elementary numerical methods. Each method will be applied to problem (1). The three methods provide successively better approximations.

We want to approximate the solution of the initial value problem

$$\frac{dy}{dx} = f(x, y), \qquad y(x_0) = y_0$$

on an interval $x_0 \leq x \leq b$. For n a positive integer, we divide the interval into n parts using points

$$x_0 < x_1 < x_2 < \cdots < x_n = b.$$

For simplicity, we assume that each part has the same width, or *step size*, $h = x_{i+1} - x_i = (b - x_0)/n$; therefore $x_i = x_0 + ih$. At each point x_i we seek an approximation, which we call y_i, to the true solution $y(x_i)$ at x_i:

$$y_i \approx y(x_i).$$

The Euler Method.   The simplest numerical solution method is due to Euler, and is based on the tangent line approximation to a function. Given the initial value y_0, we define y_i recursively by

$$y_{i+1} = y_i + h\, f(x_i, y_i), \qquad i = 0, 1, \ldots, n - 1.$$

This formula is derived as follows:

$$
\begin{aligned}
y(x_{i+1}) &\approx y(x_i) + hy'(x_i), && \text{by the tangent line approximation} \\
&= y(x_i) + hf(x_i, y(x_i)), && \text{using the differential equation} \\
&\approx y_i + hf(x_i, y_i), && \text{since } y_i \approx y(x_i) \\
&= y_{i+1}.
\end{aligned}
$$

The approximations $y(x_i) \approx y_i$ should become better and better as h is taken smaller and smaller.

EXAMPLE.   Consider the initial value problem (1):

$$\frac{dy}{dx} = \frac{x}{y}, \qquad y(0) = 1.$$

We wish to approximate $y(0.3)$ using the Euler Method with step size $h = 0.1$ and three steps. We find

$$x_0 = 0, \quad x_1 = 0.1, \quad x_2 = 0.2, \quad x_3 = 0.3$$
$$y_0 = 1$$
$$y_1 = y_0 + hf(x_0, y_0) = y_0 + hx_0/y_0 = 1$$
$$y_2 = y_1 + hx_1/y_1 = 1.01$$
$$y_3 = 1.0298.$$

Thus

$$y(0.3) = \sqrt{1.09} = 1.044030... \approx y_3 = 1.0298$$
$$\text{Error} = |y(0.3) - y_3| = 0.0142....$$

Next use $h = 0.05$ and 6 steps:

$$y_0 = 1$$
$$y_1 = 1$$
$$y_2 = 1.0025$$
$$y_3 = 1.0075$$
$$y_4 = 1.0149$$
$$y_5 = 1.0248$$
$$y_6 = 1.0370$$

$$\text{Error} = |y(0.3) - y_6| = 0.0070....$$

Note that cutting the step size in half has the effect of cutting the error approximately in half.

If the initial value problem is "sufficiently smooth", then the Euler Method is a *first order method*. That is, at each point in the interval $[x_0, b]$ we have,

$$\text{Error} \leq Ch,$$

where C is a constant that depends on $f(x, y)$, its partial derivatives, the initial condition, and the interval, but not on h. Moreover, one can prove that the error is proportional to h. (An initial value problem is sufficiently smooth if $f(x, y)$ has continuous partial derivatives of sufficiently high order. For the Euler Method we need to assume that f and its partial derivatives of orders up to two are continuous.) Thus, to gain an additional digit of accuracy, one would reduce the step size by a factor of $1/10$.

The Euler Method can be implemented using *Maple* as follows:

```
EulerMethod := proc(f, x0, y0, h, n) local a, i, x, y;
      x := evalf(x0); y := evalf(y0); a := [[x, y]];
      for i from 1 to n do
          y := y + h*f(x, y); x := x + h; a := [op(a), [x, y]];
      od; end;
```

To apply this program to the initial value problem (1), we first define the function $f(x, y) = x/y$, and then run the routine. So,

```
f := (x, y) -> x/y;
em := EulerMethod(f, 0, 1, 0.1, 10);
```

will produce a list of approximate solution values at $x = 0, 0.1, 0.2, \ldots, 1$. Note that **EulerMethod** uses a programming construct called a "for" loop. The program uses the local variable **a** to store the list of points; new points are added to the list using the *Maple* function **op** as soon as they are computed.

To approximate the solution at a point x that is not an x_i, we would have to *interpolate*. A simple way to do this is to "connect the dots" by drawing straight line segments from each point (x_i, y_i) to the next point (x_{i+1}, y_{i+1}); thus $y_a(x)$ would be the piecewise linear function connecting the computed points (x_i, y_i). To graph $y_a(x)$, type **plot(em)**.

The Improved Euler Method.   We again start with the tangent line approximation, replacing the slope $y'(x_i)$ by the average of the two slopes $y'(x_i)$ and $y'(x_{i+1})$. This yields

$$y(x_{i+1}) \approx y(x_i) + h \, \frac{y'(x_i) + y'(x_{i+1})}{2}$$

$$= y(x_i) + h \, \frac{f(x_i, y(x_i)) + f(x_{i+1}, y(x_{i+1}))}{2}.$$

We now use the approximation $y(x_i) \approx y_i$ and the previous Euler approximation $y(x_{i+1}) \approx y_i + hf(x_i, y_i)$, to get

$$y(x_{i+1}) \approx y_i + h \, \frac{f(x_i, y_i) + f(x_{i+1}, y_i + hf(x_i, y_i))}{2}$$

$$= y_i + h \, \frac{y_i' + f(x_{i+1}, y_i + hy_i')}{2}$$

$$= y_{i+1},$$

where $y_i' = f(x_i, y_i)$. Given the initial value y_0, this analysis leads to the recursive formula

$$y_{i+1} = y_i + h \, \frac{y_i' + f(x_{i+1}, y_i + hy_i')}{2}, \qquad i = 0, 1, \ldots, n - 1.$$

EXAMPLE.   We again consider problem (1) and find an approximation to $y(0.3)$ with $h = 0.1$. We obtain

$$y_1 = y_0 + h \, \frac{y_0' + f(x_1, y_0 + hy_0')}{2}$$

$$= y_0 + \frac{h}{2} \left(\frac{x_0}{y_0} + \frac{x_1}{y_0 + hx_0/y_0} \right)$$

$$= 1.005$$

$$y_2 = 1.019828$$

$$y_3 = 1.044064.$$

So Error $= |y(0.3) - y_3| = 0.000033....$ Note that the error in the Improved Euler Method with $h = 0.1$ is considerably less than the error in the Euler Method with $h = 0.05$.

For sufficiently smooth problems, the Improved Euler Method is a *second order method*; i.e., the error is proportional to h^2. Thus, to gain one more digit of accuracy in the approximation, one would reduce the step size by a factor of $1/\sqrt{10} \approx 0.32$. Equivalently, reducing the step size by a factor of $1/10$ will produce two more digits of accuracy.

The Runge-Kutta Method. Given the initial value y_0, this method is defined recursively by

$$y_{i+1} = y_i + \frac{h}{6}(k_1 + 2k_2 + 2k_3 + k_4),$$

where

$$k_1 = f(x_i, y_i)$$

$$k_2 = f(x_i + \frac{h}{2}, y_i + \frac{h}{2}k_1)$$

$$k_3 = f(x_i + \frac{h}{2}, y_i + \frac{h}{2}k_2)$$

$$k_4 = f(x_{i+1}, y_i + hk_3).$$

Note that a weighted average of the "slopes" k_1, k_2, k_3, k_4 is used in the formula for y_{i+1}.

EXAMPLE. Consider problem (1) and again find an approximation to $y(0.3)$, but now let $h = 0.3$ and take just one step. Then

$$k_1 = f(0, 1) = 0$$

$$k_2 = f(0 + 0.15, 1 + 0.15(0)) = 0.15$$

$$k_3 = f(0.15, 1 + 0.15(0.15)) = 0.146699$$

$$k_4 = f(0.3, 1 + 0.3(0.146699)) = 0.287354$$

$$y_1 = 1 + \frac{0.3}{6}(0 + 2(0.15 + 0.146699) + 0.287354)$$

$$= 1.044038 \approx y(0.3).$$

So Error $= |y(0.3) - y_1| = 0.000007....$ We see that the error in the Runge-Kutta Method with $h = 0.3$ is less than the error in the Improved Euler Method with $h = 0.1$.

For sufficiently smooth problems, the Runge-Kutta Method is a *fourth order method*; i.e., the error is proportional to h^4. Reducing the step size by a factor of $1/10$ will produce four more digits of accuracy.

We have illustrated these methods on problem (1). They can be applied in exactly the same manner to any first order equation. Furthermore, they can be applied using different step sizes at each step.

There are many numerical methods for initial value problems in addition to the ones we have presented here. The command **dsolve(..., numeric)** features each of the above methods and several additional methods. By default, **dsolve(..., numeric)** uses a method called "rkf45". Other methods can be selected with appropriate options. For example, to compute a numerical solution using the Euler Method with fixed step size, you would use the option **method = classical[foreuler]**. See the online help for **dsolve[numeric]** for a list of available methods.

The default method, rkf45, combines a fourth order method and a fifth order method developed by Fehlberg, both similar to the fourth order Runge-Kutta Method discussed above. See L. Shampine, **Numerical Solution of Ordinary Differential Equations**, Chapman & Hall, 1994 for further details. The rkf45 method uses *variable step sizes*, choosing the step size at each step to try to achieve the desired accuracy. All efficient modern numerical solvers use variable step size methods. Such methods are especially effective for calculating solutions that oscillate or are singular. The rkf45 method is suitable for a wide variety of initial value problems; we recommend its use for the problems in this text.

For some initial value problems (referred to as *stiff*), the rkf45 method is less satisfactory. Two other methods (**method = mgear** and **method = lsode**) are recommended for stiff problems. For more information on stiff problems and their numerical solution, you can consult Shampine's book (as above) or D. Kahaner, C. Moler, and S. Nash, **Numerical Methods and Software**, Prentice Hall, Inc., 1989.

The type of error discussed above is called *discretization error*. In addition, there is *round-off error*, which arises because the computer uses a fixed, finite number of digits. Letting $\tilde{y}_a(x)$ denote the actual computed approximate solution, which includes round-off error, the total error can be written

$$y(x) - \tilde{y}_a(x) = (y(x) - y_a(x)) + (y_a(x) - \tilde{y}_a(x))$$
$$= \text{Discretization Error} + \text{Round-off Error}.$$

Since most computers that run *Maple* carry 16 digits, the major portion of the error will be the discretization error. Thus, for most problems, round-off error can be safely ignored. For this reason we will not distinguish between $y_a(x)$ and $\tilde{y}_a(x)$.

REMARK. With numerical methods as with qualitative methods, we obtain information on the solution—quantitative in the one case and qualitative in the other—without the use of a solution formula.

The DEplot Command

Earlier, we explained how to use **dsolve(. . ., numeric)** and **plot** to obtain and plot numerical solutions of initial value problems. In this section we describe **DEplot**, a command designed specifically to plot numerical solutions of differential equations. It combines a plotting routine with the numerical solver **dsolve(...,numeric)**.

It also plots families of solutions corresponding to multiple initial conditions (*cf.* Chapter 8). However, one cannot readily use **DEplot** to compute the value of the numerical solution at a particular point. Recall that **dsolve(. . ., numeric)** features several numerical methods. The default method used by **DEplot** is the fourth order Runge-Kutta method with fixed step size. We recommend, however, that you use **DEplot** with **method = rkf45**; this choice needs to be specified explicitly, as in the following example.

Consider the initial value problem

$$y' = x - 1 + \cos y, \quad y(0) = 1. \tag{3}$$

To plot a numerical solution of this initial value problem on the interval $[0, 2]$, we type

```
with(DEtools):
DEplot(diff(y(x), x) = x − 1 + cos(y(x)), y(x), x = 0..2, {[y(0) = 1]},
    method = rkf45, arrows = NONE, linecolor = black);
```

The result is shown in Figure 2.

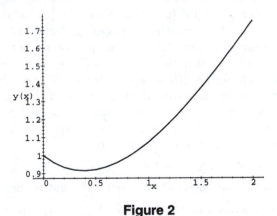

Figure 2

The first argument to **DEplot** specifies the differential equation; the second argument specifies the dependent variable and the independent variable; the third argument specifies the interval on which to plot the solution; the fourth argument specifies one or more initial conditions; the fifth argument specifies the numerical method; and the sixth argument specifies that only the solution, and not the direction field, is plotted. If this last argument is omitted, both the direction field and the solution are plotted.

EXERCISE. Solve the initial value problem (3) with **dsolve(. . . , numeric)**, and graph the solution on the interval $[0, 2]$ using **plot**. Display the graphs produced by **plot** and **DEplot** on the same plot. Are the two graphs distinguishable?

DEplot evaluates the numerical solution $y(x)$ at 21 equally spaced x-values in the interval $[a, b]$ over which we are plotting (with distance $h = (b - a)/20$ between adjacent values), and then connects the points $(x, y(x))$ with straight line segments. To obtain a smoother graph, use the option **stepsize = h**, where h is a positive number smaller than $(b - a)/20$. Note that this **stepsize** is different from the step size used in the numerical method.

EXERCISE. Graph the solution of the initial value problem (2) with **DEplot**; in particular, find a value for x^*. You will find it useful to specify the range with the option $y = 0..z$, for some positive z, say $z = 10$.

REMARK. *Maple* also contains the command **odeplot**, which plots numerical solutions of differential equations. We do not feel that **odeplot** offers any advantages over **plot** and **DEplot**, and thus it will not be discussed in this book.

Controlling the Error in dsolve(. . . , numeric)

The numerical solver **dsolve(. . . , numeric)** attempts to produce an approximate solution $y_a(x)$ satisfying the accuracy requirement

$$\text{Error} = |y(x) - y_a(x)| \text{ is approximately} \leq 10^{-8}.$$

A more accurate approximate solution can be obtained by redefining the global variable **Digits**; for example by typing

   Digits := 17;

Then

   sol3 :=dsolve(ivp, y(x), numeric, maxfun = 2000, startinit = TRUE);
   ya3 := u –> subs(sol3(u), y(x));

produces a more accurate approximation to the solution of (1): **ya3(1.5)** equals 1.802775637731997, which is a 15-digit approximation to the exact value $\sqrt{13}/2$. (Note that the desired accuracy cannot be achieved with **maxfun = 1000**.)

The value of **Digits** determines the number of digits that *Maple* uses in its calculations and also determines the accuracy that **dsolve(. . . , numeric)** attempts to achieve. Specifically, **dsolve(. . . , numeric)** attempts to control absolute error according to

$$\text{Absolute Error} = |y(x) - y_a(x)| \text{ is approximately} \leq 10^{2-\text{Digits}}$$

and relative error according to

$$\text{Relative Error} = \frac{|y(x) - y_a(x)|}{|y(x)|} \text{ is approximately } \leq 10^{2-\text{Digits}}.$$

The default value for **Digits** is 10, so the default option leads to approximately 8-digit accuracy, as indicated above (but see the next section).

Since **Digits** is a *global variable*, changing its value affects the number of digits carried along in every floating point calculation, including the results of **evalf**.

REMARK. Increased accuracy is achieved in the same way when using **DEplot** with **method = rkf45**; *i.e.*, by setting **Digits** equal to an integer greater than 10.

Reliability of Numerical Methods

We have claimed that **dsolve(..., numeric)**, when used with the default options, leads to approximately 8-digit accuracy. More precisely, the default options ensure that the *local error*, *i.e.*, the error made in a step of length h, is approximately $10^{-8} \times h$. One is generally more interested in the *global error*, *i.e.*, the cumulative error committed in taking the necessary number of steps to get from the initial point x_0 to some target point x. This error cannot be controlled completely by the numerical method because it depends on the differential equation as well as on the numerical method. The key issue is whether the differential equation is stable or unstable; these terms were defined in Chapter 5.

We have seen equations for which **dsolve(..., numeric)** gives accurate results. Now we look at an example where **dsolve(..., numeric)** gives poor results.

EXAMPLE. Consider the initial value problem

$$\frac{dy}{dx} - y = -3e^{-2x}, \qquad y(0) = 1.$$

The exact solution is $y(x) = e^{-2x}$; we plot it in Figure 3.

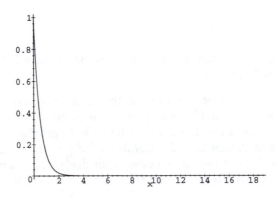

Figure 3

Next we find the numerical solution using **dsolve(..., numeric)**, and plot it together with the exact solution:

```
exact4 := plot(exp(-2*x), x = 0..19):
exact4;
ivp4 := {diff(y(x), x) = y(x) -3*exp(-2*x), y(0) = 1}:
sol4 := dsolve(ivp4, y(x), numeric, maxfun = 1000, startinit = TRUE):
ya4 := u -> subs(sol4(u), y(x)):
appr4 := plot(ya4, 0..19):
display({appr4, exact4});
```

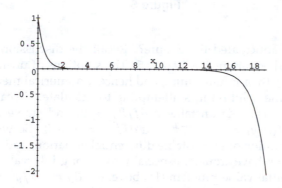

Figure 4

We see from Figure 4 that the graphs are indistinguishable from 0 to about 14, but then the curves separate sharply. We find that $y_a(19) = -2.05\ldots$, which is very different from the exact value $y(19) = e^{-38} \approx 0.314 \times 10^{-16}$.

How can this failure be explained? The solution to our differential equation with initial condition $y(0) = 1 + \epsilon$ is $y(x) = e^{-2x} + \epsilon e^x$. Figure 5 shows a plot of the solutions corresponding to initial values $1 - (0.5 \times 10^{-7}), 1 - (0.25 \times 10^{-7}), 1, 1 + (0.25 \times 10^{-7}), 1 + (0.5 \times 10^{-7})$.

We see that solutions with initial values slightly different from 1 separate very sharply from the solution with initial value exactly 1 after about 12; those with initial values greater than 1 approach $+\infty$, and those with initial values less than 1 approach $-\infty$. Now **dsolve(..., numeric)**, like any numerical method, introduces errors. These errors—whether discretization errors or round-off errors—have caused the numerical solution to jump to a solution that started just below 1. Once on such a solution, the numerical solution will follow it or another such solution. Since these solutions approach $-\infty$, the numerical solution does likewise.

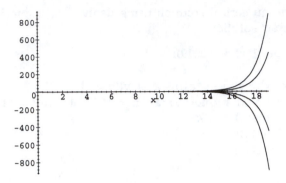

Figure 5

Could we have anticipated this failure? Recall the discussion of stable and unstable differential equations in Chapter 5. The solutions of unstable equations are very sensitive to their initial values, and hence a numerical method will have trouble following the solution it is attempting to calculate. Recall also that an equation $dy/dx = f(x, y)$ is unstable if $\partial f/\partial y > 0$, and stable if $\partial f/\partial y \leq 0$. For our example, $f(x, y) = y - 3e^{-2x}$ and $\partial f/\partial y = 1 > 0$. So we are trying to approximate the solution of an unstable differential equation, and we should not be surprised that we have trouble, especially over long intervals. (By contrast, note that for the initial value problem (1) above, $\partial f/\partial y = -x/y^2 < 0$ for $x > 0$.) All this suggests caution when dealing with unstable equations. In fact, caution is always recommended when modeling physical problems with unstable equations, especially over long intervals (*cf.* the discussion in Chapter 5).

In assessing the reliability of a numerically computed solution there is another test one can make. Suppose we have made a calculation using the default options, as we have done with our example. Then one can do another calculation, using a more stringent accuracy requirement, and compare the results. If they are nearly the same, one can have reasonable confidence that they are both accurate; but if they differ substantially, then one should suspect that the first calculation is not accurate.

Let's do this with our example:

```
Digits := 11;
sol5 := dsolve(ivp4, y(x), numeric, maxfun = 1000, startinit = TRUE):
ya5 := u -> subs(sol5(u), y(x)):
appr5 := plot(ya5, 0..19);
```

Next we plot both numerical solutions together with the exact solution:

```
display({exact4, appr4, appr5});
```

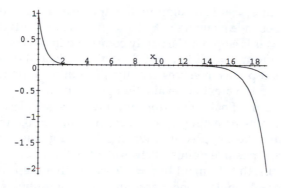

Figure 6

We see from Figure 6 that the second numerical solution is accurate over a longer interval, out to about 16, whereas the first was accurate out to about 14. In particular, the two numerical solutions differ after about 14. Even if we didn't know the exact solution, we could have concluded that the first numerical solution is accurate out to, but not beyond, about 14.

In summary, we have learned that local accuracy—the accuracy that a numerical method can control—leads to reasonable global accuracy if the differential equation is stable, and that the reliability of a numerical solution can also be checked by computing another numerical solution with a more stringent accuracy requirement.

REMARK. The comments we have made about the reliability of **dsolve(..., numeric)** apply of course to **DEplot** when used with **method = rkf45**.

Further Remarks on dsolve(...,numeric), plot and DEplot

We begin by explaining why we have included the option **startinit = TRUE** in **dsolve(..., numeric)**. Consider the initial value problem

$$\frac{dy}{dx} = 1 - 2xy, \quad y(0) = 0, \tag{4}$$

and its numerical solution with **dsolve(..., numeric)**, but without the option **startinit = TRUE** or, equivalently, with the option **startinit = FALSE**. This is done with the commands

> **ivp6 := {diff(y(x), x) = 1 − 2*x*y(x), y(0) = 0}:**
> **sol6 := dsolve(ivp6, y(x), numeric, maxfun = 2000):**
> **ya6 := u −> subs(sol6(u), y(x)):**

If we evaluate **ya6** at some point, say $x = 0.1$, the procedure **sol6** produced by **dsolve(..., numeric)** starts at $x = 0.0$, employs the initial condition $y(0) = 0$, and steps from $x = 0.0$ to $x = 0.1$ to compute the approximate value $0.0993\ldots$ for **ya6(0.1)**. Next suppose we evaluate **ya6** at $x = 0.2$. Then **dsolve(..., numeric)** starts at the point $x = 0.1$, employs the newly computed initial condition $y(0.1) = 0.0993\ldots$, and steps from $x = 0.1$ to $x = 0.2$. With the default option, when **ya6** is evaluated at several points in succession, the procedure starts from the closer of the initial point and the previous evaluation point. By starting at $x = 0.1$ and using as initial value the final value from the previous calculation, **dsolve(..., numeric)** sensibly avoids repeating the calculation from $x = 0.0$ to $x = 0.1$. The value of this feature is best appreciated when calculating solution values at an increasing sequence of positive points, such as $x = 0.1, 0.2, \ldots, 0.9, 1.0$. One steps from $x = 0.0$ to $x = 0.1$, then from 0.1 to $x = 0.2$, *etc.*, without repeating any of the previous calculations. A similar benefit accrues when computing solution values at a decreasing sequence of negative points.

One must beware, however, that computing solution values at a sequence of points that is neither increasing nor decreasing from the initial data point can lead to inaccurate results. To see this, repeat the *Maple* command defining **sol6**, and evaluate **ya6** at $x = 2.6$ to get $ya6(2.6) = 0.212\ldots$. Repeat the *Maple* command defining **sol6**, and then evaluate **ya6**, first at $x = 5.0$ and then at $x = 2.6$, to get $-0.291\ldots$, which is in sharp conflict with the value $ya6(2.6) = 0.212\ldots$ obtained earlier. How can this discrepancy be explained? When $ya6(2.6) = 0.212\ldots$ is computed the first time, **dsolve(..., numeric)** starts at the point $x = 0.0$, and steps across to $x = 2.6$. When $ya6(2.6)$ is computed the second time, **dsolve(..., numeric)** starts at the point $x = 0.0$, and steps across to $x = 5.0$ to get $ya6(5.0) = 0.102\ldots$. Then, **dsolve(..., numeric)** starts at the point $x = 5.0$, employs the initial condition $y(5.0) = 0.102\ldots$, and steps back from $x = 5.0$ to $x = 2.6$. If our numerical method tracked the solution exactly, we would have gotten the value $.212\ldots$. Of course, the method is not exact; it includes discretization and round-off errors. Since our differential equation is unstable to the left, these errors are severe. (When going to the left, the Theorem on p. 47 of Chapter 5 applies with the roles of K and L reversed. In particular, an equation is unstable to the left if $\partial f/\partial y < 0$. In the case at hand, $\partial(1 - 2xy)/\partial y = -2x < 0$. Thus the numerical calculation from 5.0 to 2.6 is inaccurate.)

The option **startinit = TRUE** ensures that each calculation starts at the original initial data point, and thus avoids the problem discussed in the previous paragraph. If **ya6** is defined with this option, and if you calculate $ya6(2.6)$ in the two ways discusssed above, you will get the same value. So we recommend using the option **startinit = TRUE** unless you are computing values at a sequence that either is increasing or is decreasing from the initial data point.

A similar problem arises when using **plot** to graph a numerical solution obtained with **dsolve(..., numeric)** without the option **startinit = TRUE**. When applied to a function y, the **plot** command calculates $y(x)$ at a set of x-values, and then connects the points $(x, y(x))$ by straight line segments. It chooses the x-values at which to evaluate $y(x)$, choosing fewer points where $y(x)$ is close to a straight line and more where it is not close to a straight line; for this reason it is called an adaptive plotter.

Thus the number and order of the points **plot** chooses is unpredictable. Using **startinit = TRUE** avoids the problem mentioned above.

EXERCISE. Graph the solution to the initial value problem (4) over the interval $[-5, 5]$ first with and then without the option **startinit = TRUE**.

There is one potential drawback: plotting with **startinit = TRUE** may be slow. In certain situations, however, plotting may be speeded up by using the option **adaptive = false** with **plot**. Suppose the initial data point is at the left end of the interval. With **adaptive = false**, **plot** evaluates the numerical solution at a sequence of points increasing from the initial data point. You can thus safely omit the option **startinit = TRUE**, and thereby speed up the **plot**.

Next suppose you wish to plot **ya6** over the interval $[-5, 0]$. Even with **adaptive = false**, the **plot** command may yield an inaccurate result. For it first computes **ya6** at -5, and then at a sequence of increasing points, getting inaccurate values. This problem can be overcome as follows:

> **plot(ya6, adaptive = false, sample = [seq(−0.1*i, i = 0..50)]);**

The option **sample = [seq(−0.1*i, i = 0..50)]**, used in conjunction with the option **adaptive = false**, forces **plot** to evaluate **ya6** at the sequence $0.0, -0.1, -0.2, ..., -5$.

EXERCISE. Plot the solution of (4) over $[-5, 0]$, using **dsolve(. . ., numeric)** without **startinit = TRUE**, first with

> **plot(ya6, adaptive = false);**

and then with

> **plot(ya6, adaptive = false, sample = [seq(−0.1*i, i = 0..50)]);**

The problem we have been discussing doesn't arise with the use of **DEplot**. If we are plotting over an interval and the initial data point is the left end point of the interval, **DEplot** evaluates the numerical solution at a sequence of points increasing from the initial data point; if the initial data point is the right end point of the interval, the numerical solution is evaluated at a sequence of points decreasing from the initial data point; and if the initial data point is in the interior of the interval, **DEplot** first plots to the right of the initial data point and then to the left.

Chapter 8

Features of *Maple*

This chapter is a continuation of Chapter 3, *Doing Mathematics with Maple*. We describe some of the more advanced features of *Maple*, focusing on the *Maple* commands and techniques most useful for studying differential equations. We also describe some aspects of *Maple*'s internal structure that should help in working with complicated expressions and commands.

Names and Values.   A *Maple* object usually has both a *name* and a *value*. For example, the symbol x has the name "x". Initially, its value is also its name, "x". However, if you type

> x := 2;

$$x := 2$$

the situation changes. The name of x is still "x", but its value is now 2.

When you enter an input statement in a *Maple* session, *Maple* generally tries to find values for all of the variables in the statement. *Maple* substitutes the value of a variable for its name when it processes the input. If the value of a variable is another variable, then *Maple* will in turn substitute the value of that variable, and so on. This is called *full evaluation*.

Note that once a value has been assigned to a variable it ceases to be a variable in the true sense of the word, and becomes instead just a name for its value. In certain situations, you may want to insist that *Maple* treat something as a variable. You can do this by enclosing the variable name in right quotes ('). For example:

> 'x'^2;

$$x^2$$

Compare this with the unquoted version:

> x^2;

$$4$$

The effect of evaluation on a quoted name is to strip away one set of right quotes. No further evaluation is performed. We will see several examples below where

83

right quotes are useful to prevent full evaluation.

Clearing Values.   One of the most common sources of errors in using *Maple* is the failure to keep track of variable or function definitions. For example, suppose that early in a session you type
>   i := 3;

$$i := 3$$

You keep working and twenty minutes later you want to evaluate the sum $1 + 1/2 + \ldots + 1/9$. So, you type
>   **sum(1/i, i = 1..9);**
    Error, (in sum) summation variable previously assigned,
       second argument evaluates to, 3 = 1 ..   9
Because i already had a value, *Maple* saw this command as **sum(1/3, 3 = 1..9)**, and objected to it.

One way to avoid this kind of error is by using right quotes to suppress evaluation. For example,
>   **sum(1/'i', 'i' = 1..9);**

$$\frac{7129}{2520}$$

A simpler way to avoid the error is to clear the value of i (by typing **i := 'i':**) before you use it. Recall that the **restart** command will clear all the values in a *Maple* session.

Functions and Expressions.   *Maple* distinguishes between a *function* and an *expression*. In order to use *Maple* effectively, you must learn to make the same distinction. You should review the section *User-defined Functions and Expressions* in Chapter 3 if you've forgotten the difference between functions and expressions.

There are several ways to define a function in *Maple*. The simplest way is to use *Maple's* arrow notation.
>   **restart;**
>   **f := x -> x^2 - 1;**

$$f := x \rightarrow x^2 - 1$$

>   **f(2);**

$$3$$

Functions of several variables can also be defined.
>   **phi := (x, y) -> x^2 + y^2;**

$$\phi := (x, y) \rightarrow x^2 + y^2$$

Note that the list of arguments must be enclosed in parentheses.

A function can be evaluated with a symbolic argument as well as a numeric argument:

> **f(y);**

$$y^2 - 1$$

The result of this evaluation is the expression $y^2 - 1$. This is a subtle point: the symbol **f** is the name of a *function*; the symbol **f(x)** is the name of an *expression* involving **x**.

To remove the definition of a function you must unassign the name of the function.

> **f := 'f':**
> **f(y);**

$$f(y)$$

The command **unapply** can be used to turn an expression into a function. For example, suppose we define

> **expr1 := sin(x) + cos(x);**

$$expr1 := \sin(x) + \cos(x)$$

We then type

> **g := unapply(expr1, x);**

$$g := x \rightarrow \sin(x) + \cos(x)$$

The command **unapply** takes the first argument, which must be an expression (or a name of an expression) containing one or more variables, evaluates this expression, and then makes it into a function of the specified variables. To make a function of two variables, you could type, for example,

> **expr2 := x^2 + y^2;**

$$expr2 := x^2 + y^2$$

> **h := unapply(expr2, x, y);**

$$h := (x, y) \rightarrow x^2 + y^2$$

> **h(2, 3);**

$$13$$

There is a subtle difference between using **unapply** and the arrow notation to define a function. Note that in the examples of **unapply**, the names $expr1$ and $expr2$ were evaluated and replaced with their values. This does not happen if you define the function with an arrow.

```
> g := x -> expr1;
```

$$g := x \rightarrow expr1$$

```
> g(0);
```

$$\sin(x) + \cos(x)$$

What happens here is that when *Maple* evaluates a function defined with the arrow notation, it *first* substitutes the argument (in this case x) into the expression on the right-hand side, and *then* evaluates the right-hand side. Since x does not appear as a variable on the right-hand side in this example, the function we define this way is actually independent of x. Thus g has been defined to be a function that returns as its *constant* value the expression $expr1$. The point of this discussion is that the arrow notation can only be used when the argument (in this case, x) appears explicitly on the right-hand side of the arrow. The command **unapply** should be used when you want to build a function using a previously defined expression.

A third way of defining a function is by creating a *Maple procedure* using the command **proc**. This involves some understanding of programming. For an introduction to *Maple* programming, see, for example, B. Char, *et al.*, **First Leaves: A Tutorial Introduction to Maple V**, Springer-Verlag, 1992. An example of a simple *Maple* program is given in the glossary of this book.

It is possible in *Maple* to define values of functions at particular points. Suppose, for example, that we want to define the function $f(x) = \sin(x)/x$. We know from elementary calculus that the limit of this function as x approaches 0 is 1, but if we evaluate the function at $x = 0$, *Maple* gives an error. We could circumvent this problem by doing the following:

```
> f := x -> sin(x)/x;
```

$$f := x \rightarrow sin(x)/x$$

```
> f(0) := 1;
```

$$f(0) := 1$$

What we have done here is merely to specify that at the point $x = 0$, *Maple* should ignore the general definition $f := x \rightarrow \sin(x)/x$, and instead just set $f(0) = 1$. This feature is occasionally useful when working with functions that have expressions in the denominator that might take on zero values. However this syntax should NEVER be used to define a general function. For example,

```
> f := 'f':
> f(x) := sin(x)/x;
```

$$f(x) := \frac{sin(x)}{x}$$

> **f(x);**

$$\frac{sin(x)}{x}$$

> **f(1);**

$$f(1);$$

In this example, $f(x)$ is merely a name for the expression $sin(x)/x$. There is no substitution of values for x. The symbol $f(1)$ has *not* been defined, so *Maple* just echoes what we've typed.

If you accidentally type a definition like **f(x) := sin(x)/x**, you can clear it by reassigning the name of the function.

> **f := 'f';**

$$f := f$$

Substitution. We have seen that a function can be evaluated at various values of the argument. A similar effect can be accomplished with expressions using the substitution command, **subs**.

> **subs(x = 5, x^2 − 1);**

$$24$$

The **subs** command uses one or more equations to substitute for variables in an expression (the last argument of the command). Here is another example:

> **subs(x = a, y = b, z = c, (x^2 − 3*y)/z);**

$$\frac{a^2 - 3b}{c}$$

When you use the **subs** command with more than one equation (as in the preceding example), the substitution is done from left to right. Thus in the preceding example a is first substituted for x, then b for y, then c for z. If you want the substitutions to occur simultaneously, you must enclose the equations in braces. Here is an example in which we switch x and y in an expression.

> **subs({x = y, y = x}, x/y);**

$$\frac{y}{x}$$

You can test your understanding of the **subs** command by predicting what the output of the preceding statement would be if the braces were omitted.

The most common use of **subs** is to manipulate expressions that have been previously typed in by the user or generated by *Maple*. Here is an example in which we find the solution of the differential equation $y' = xy$ with initial condition $y(0) = 1$. First we find the general solution.

> **soln := dsolve(diff(y(x), x) = x*y(x), y(x));**

$$soln := y(x) = e^{\left(\frac{1}{2}x^2\right)} \_C1$$

By inspecting the solution, we see that the undetermined constant _C1 must be 1 in order for the solution to satisfy the initial condition. Thus the solution is obtained by substituting 1 for _C1.

> **subs(_C1 = 1, soln);**

$$y(x) = e^{\left(\frac{1}{2}x^2\right)}$$

Of course, we could also have specified the initial condition as an argument to **dsolve** in this example.

Equations vs. Assignments.   In *Maple*, an *equation* is different from an *assignment*. For example, $x = 2$ is an equation, whereas $x := 2$ is an assignment. Compare the effects of entering an equation and an assignment.

> **x = 2;**

$$x = 2$$

> **x;**

$$x$$

> **x := 2;**

$$x := 2$$

> **x;**

$$2$$

> **x := 'x':**

Note that an equation has no effect on the value of the variables in the equation. Equations can even be assigned names.

> **nice_eqn := E = m*c^2;**

$$nice\_eqn := E = m\,c^2$$

The right- and left-hand sides of an equation can be extracted using the functions **rhs** and **lhs**.

> **lhs(nice_eqn) − rhs(nice_eqn);**

$$E - m\,c^2$$

Differentiation.   The distinction between functions and expressions becomes important when differentiating.   *Maple* has two commands for differentiation. The **diff** command operates on expressions.

```
> f := t -> t^2:
> diff(f(t), t);
```

$$2t$$

Notice that the output of **diff** is another expression. Combining **diff** with the sequence operator, as in the example **diff(f(t), t$2)**, you can compute second (or higher order) derivatives.

The **D** operator differentiates functions, and returns a function as its output.

```
> D(f);
```

$$t \rightarrow 2t$$

When you differentiate an expression, you must state explicitly which variable should be used. When you differentiate a function, you must not mention the name of the variable.

One place where you must use the **D** command is to specify initial values for derivatives. In order to solve the second order initial value problem

$$y'' = 9y + 2x, \qquad y(0) = 1, \qquad y'(0) = 0,$$

you could type

```
> de := diff(y(x), x$2) = 9*y(x) + 2*x;
```

$$de := \frac{\partial^2}{\partial x^2} y(x) = 9y(x) + 2x$$

```
> initial := y(0) = 1, D(y)(0) = 0;
```

$$initial := y(0) = 1, D(y)(0) = 0$$

```
> dsolve({de, initial}, y(x));
```

$$y(x) = -\frac{2}{9}x + \frac{29}{54}e^{(3x)} + \frac{25}{54}e^{(-3x)}$$

The notation for higher derivatives is more complex. For example, suppose you want to specify the initial value $y'''(0) = 1$. One way to write this condition is

```
> D(D(D(y)))(0) = 1;
```

$$(D^{(3)})(y)(0) = 1$$

As you might imagine, all those layers of **D**'s and parentheses can become confusing. *Maple* offers an alternative:

```
> (D@@3)(y)(0) = 1;
```

$$(D^{(3)})(y)(0) = 1$$

Type **?@** for more information about this operator.

More About Sequences, Sets and Lists.   Sequences, sets, and lists are useful when you want to do many things with a single command. For example, to plot the expressions $x^2 + c$ for values of c ranging from -5 to 5, you could type **plot({x^2 + c \$ c = −5..5}, x = 0..5)**. The syntax **{x^2 + c \$ c = −5..5}** creates a *set* of functions, which are then plotted by the **plot** command.

An important difference between the sequence operator and **seq** command is the order of evaluation. The sequence operator first evaluates the expression to the left of the dollar sign, and then substitutes the value of the index into the expression. The **seq** command first substitutes the value of the index and then evaluates the expression. This is a rather subtle difference, but it can have serious consequences. Consider the following commands.

> **solve(x^2 = i, x) \$ i = 1..2;**

    Error, wrong number (or type) of parameters in function \$

> **seq(solve(x^2 = i, x), i = 1..2);**

$$1, -1, \sqrt{2}, -\sqrt{2}$$

In both input lines, we've attempted to create a sequence consisting of the square roots of the integers 1 and 2. The first attempt using the sequence operator fails because of premature evaluation; *Maple* tries to solve the equation $x^2 = i$ *before* it assigns a value to i. The second attempt works because *Maple* first assigns values to i and then evaluates the **solve** command. You could rectify the problem with the sequence operator by enclosing the expression to the left-hand side of the dollar sign in right quotes to delay evaluation. Problems with premature evaluation are likely to occur when you use the sequence operator to construct sequences out of complex expressions. In these situations the **seq** command is preferred.

There are two other reasons you might prefer the **seq** command. First, it is more flexible. Its second argument need not be a range of values; it can be a list or a set. For example, if you want to evaluate $x^2 - x$ at the points 0.13, 2, 3, and 7.4, you could type

> **list1 := [0.13, 2, 3, 7.4]:**
> **seq(x^2 − x, x = list1);**

$$-.1131, 2, 6, 47.36$$

Note that in this example we have used a *list* of parameters. You can also use a *set* of parameters, as in **x = {0.13, 2, 3, 7.4}**. Since sets are unordered, the sequence that *Maple* returns may not be in the same order as the set of parameters.

The second reason to use **seq** is that it is better suited for nested constructions used to define sequences that depend on two or more parameters. Here is an example in which we construct the set of all fractions with numerator and denominator between 1 and 3.

> **set1 := {seq(seq(a/b, b = 1..3), a = 1..3)};**

$$set1 := \{1, 2, 3, 1/2, 1/3, 2/3, 3/2\}$$

To do this with the sequence operator, it is necessary to use right quotes to prevent premature evaluation

> a := 'a':
> b := 'b':
> {'a/b $ b = 1..3' $ a = 1..3};

$$\{1, 2, 3, 1/2, 1/3, 2/3, 3/2\}$$

Elements of a set or list can be extracted with the command **op** (for "operand"). The output is in the form of a sequence.

> op(list1);

$$.13, 2, 3, 7.4$$

> op(set1);

$$1, 2, 3, 1/2, 1/3, 2/3, 3/2$$

Using **op**, we can build new lists out of old lists.

> list2 := [op(list1), a, b, c];

$$[.13, 2, 3, 7.4, a, b, c]$$

For sets, we can use the more natural command **union**.

> set1 union {a, b, c};

$$\{1, 2, 3, 1/2, c, a, b, 1/3, 2/3, 3/2\}$$

Sublists and elements of lists can be extracted using the same procedure as for sequences.

> list2[5];

$$a$$

> list2[4..7];

$$[7.4, a, b, c]$$

In fact, the same procedure also works for sets, but is inadvisable since sets are arbitrarily ordered.

The *Maple* command **map** can be used to apply a function to the elements of a list.

> map(log, list1);

$$[-2.040220829, ln(2), ln(3), 2.001480000]$$

> evalf(", 5);

$$[-2.0402, .69315, 1.0986, 2.0015]$$

The **map** command provides a way to print a *Maple* list in tabular form. In the following example, we create a "list of lists" and then print it in tabular form.

> t := [[x = j, y = sin(j)] $ j = 1..4];

$$t := [[x = 1, y = sin(1)], [x = 2, y = sin(2)], [x = 3, y = sin(3)],$$
$$[x = 4, y = sin(4)]]$$

We'd really like to see decimal approximations of these values, so we'll use **evalf** to get decimal approximations to 4 places.

> t := evalf(t, 4);

$$t := [[x = 1., y = .8415], [x = 2., y = .9093], [x = 3., y = .1411],$$
$$[x = 4., y = -.7568]]$$

> map(print, t):

$$[x = 1., y = .8415]$$

$$[x = 2., y = .9093]$$

$$[x = 3., y = .1411]$$

$$[x = 4., y = -.7568]$$

The colon at the end of the **map** command prevents *Maple* from printing an extra line containing an empty list. You could actually combine the three steps above into the single complex command **map(print, evalf([[x = j, y = sin(j)] $ j = 1..4]))**. A similar effect can be obtained with the **array** command.

> array(t);

$$\begin{bmatrix} x = 1., & y = .8415 \\ x = 2., & y = .9093 \\ x = 3., & y = .1411 \\ x = 4., & y = -.7568 \end{bmatrix}$$

We showed above how to use **op** to strip brackets or braces off a list or set. This command can also be used together with the **map** command to strip off extra brackets or braces inside a list or set.

> nestedlist := [[a, b], [c, d]];

$$nestedlist := [[a, b], [c, d]]$$

> map(op, nestedlist);

$$[a, b, c, d]$$

Compare this with the effect of the command **op(nestedlist)**.

Graphics. We will encounter several plotting commands in this course, including **plot, contourplot, implicitplot, textplot, display, dfieldplot** and **DEplot**. We briefly describe these commands in this section.

The most important thing to keep in mind with various plotting commands is the type of *Maple* object the command is designed to plot. For example, the **plot** command plots functions, expressions, parametric functions or lists of points. On the other hand, **implicitplot** is designed to plot equations (although it will plot an expression or function by turning it into an equation of the form $expression = 0$). Moreover, the distinction between functions and expressions is important in all the plotting commands. If *Maple* gives you an error message or an implausible plot when you enter a plot command, the first thing you should check is whether you are plotting the right kind of object for that particular command, and whether you are using precisely the right syntax for that object and that command.

The simplest and most versatile plotting command is **plot**. We explained in Chapter 3 how to use it to plot a single function or expression. To plot multiple functions or expressions, the collection must be specified in the form of a set or list (but not a sequence). The vertical range of the plot may be specified as an option. For example, to plot the two functions $e^x \cos x$ and $x^3 - 3x + 2$ on the interval $[-3, 3]$ with range $[-10, 10]$, we type

> **plot({exp(x)*cos(x), x^3 − 3*x + 2}, x = −3..3, y = −10..10);**

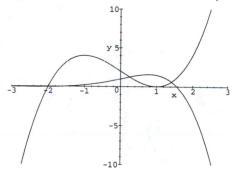

In this construction the **y** in **y = −10..10** is really just a place holder. It appears in the plot as a label on the vertical axis. Any string could appear in its place. Thus if the functions in the previous input represented voltage, you might enter the command as **plot({exp(x)*cos(x), x^3 − 3*x + 2}, x = −3..3, voltage = −10..10)**. You could also omit the label altogether and simply enter the range of the dependent variable as **−10..10**.

Plots can be assigned names and later recalled. For example,

> **plot1 := plot({exp(x)*cos(x), x^3 − 3*x + 2}, x = −3..3, y = −10..10):**

Because of the colon at the end of the command, *Maple* does not print any output. In particular, it does not display the graph. If you use a semicolon in this construction, then it *still* doesn't display the graph; instead, it prints a very long list of points in your Worksheet. (This list is the underlying data structure of the plot.)

To view the named plot, you simply type its name (in this case, **plot1**), followed by a semicolon.

The **plot** command can be used to plot curves defined by parametric equations.

> **plot([exp(−t/100)*cos(t), exp(−t/100)*sin(t), t = 0..200]);**

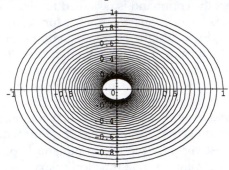

The **plot** command can also be used to plot a list of points. The list may be given explicitly or it may be generated by another command. By default, *Maple* connects the points with straight line segments.

> **plot2 := plot([[−3, 1], [−1, 2], [0.5, 8], [2, 3], [3, 2]]):**
> **plot2;**

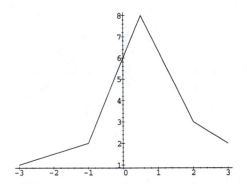

Several previously named plots can be displayed on a single graph using the **display** command. This command is in the **plots** package and must be loaded explicitly. For example, the command **display({plot1, plot2})** displays *plot1* and *plot2* on the same graph. Note that the names of the plots may appear as a set or a list, but not a sequence.

One use for the **display** command is to add textual annotations to a graph. After creating and naming a plot, you can add labels with the **textplot** command (also in the **plots** package). The argument to **textplot** is a list or set of items, each of the form **[x0, y0, `label`]**. The label is placed at coordinates $(x0, y0)$. Note that the label must be enclosed in left quotes (type **?`** for more information). For example, you could type

> **with(plots):**

> **plot3 := textplot([[1, 8, `f(x)`], [2.6, 7, `g(x)`], [1.8, −5, `h(x)`]]):**

> **display({plot1, plot2, plot3});**

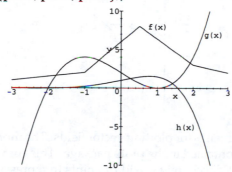

The **contourplot** command plots level curves of a function or expression; *i.e.,* sets of points on which the function or expression is constant. There are four different calling sequences for **contourplot**, but the most common usage is with an expression involving exactly two variables. For example,

> **contourplot(x^2 + y^2, x = −5..5, y = −5..5, axes = BOXED);**

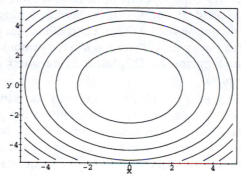

The **implicitplot** command is similar to the **contourplot** command in that it is used to plot a *single* level set of a function or expression. It is most commonly used to study functions that are defined implicitly by an equation. The important thing to remember about this command is that it takes for its argument an *equation* involving exactly two variables. (The argument can also be an expression involving two variables, in which case *Maple* uses the equation "expression=0".) You must explicitly specify the range of each of these variables. For example, you could type

> **implicitplot(y^2 + y = x^3 − x, x = −2..3, y = −3..3);**

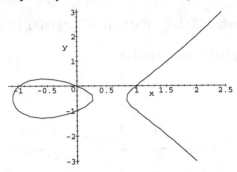

Maple has two commands for plotting vector fields. The more general command is **fieldplot**, which is contained in the **plots** package. This command places arrows of different length and direction at a grid of points to represent a vector field (see Chapter 13). The other command is **dfieldplot**, which was introduced in Chapter 6. It is in the **DEtools** package, and is designed specifically for plotting the direction fields associated with a first order ordinary differential equation. The principal difference between it and **fieldplot** is that **dfieldplot** plots vectors (or line segments) of fixed length, while **fieldplot** plots vectors whose length is proportional to the magnitude of the functions defining the vector field.

All the plotting commands let you control the look of a plot by changing certain options. On some systems, some of these options can be changed interactively using the mouse. The other options must be set before the plot is created. Type **?plot[options]** for a list of options to the main plotting commands, and **?DEtools[options]** for the options to the **DEplot** and **dfieldplot** commands. When preparing graphs for printing, you will want to use the options **color = black** (for plots of functions and for the arrows in **DEplot**) and **linecolor = black** (for solution curves plotted with **DEplot**).

The **plot** command is an adaptive plotting routine; it starts by plotting a certain number of evenly spaced points, and in regions where the graph is changing rapidly, it goes back and tries to plot more points to get a better picture. The option **numpoints** tells *Maple* how many points to start with; the default value is 49. In certain situations you may need to set a higher value of numpoints to get a smoother plot.

Plotting Families of Numerical Solutions of ODEs. In Chapter 5, we showed how to create and plot a *family* of symbolic solutions of a differential equation depending on the initial value. It is possible to do the same thing with numerical solutions. The process is different, however, because the numerical solver only works with an actual numerical initial value.

Here is an example. Consider the differential equation

> **eqn1 := diff(y(x), x) = sqrt(y(x) + x);**

$$eqn1 := \frac{\partial}{\partial x} y(x) = \sqrt{y(x) + x}$$

When we used **dsolve**, we created a set of solution curves and then plotted the set with the **plot** command. If we were to try to do this with **dsolve(. . ., numeric)**, we would have to plot a single solution at a time, and then use **display** to put all the plots on a single graph. But it is much simpler just to use **DEplot**.

> **with(DEtools):**
> **iniset := {seq([y(1) = 0.5*i], i = −2..2)};**

$$iniset = \{[y(1) = -1.0], [y(1) = -.5], [y(1) = 0], [y(1) = .5], [y(1) = 1.0]\}$$

> **DEplot(eqn1, y(x), x = 1..3, iniset, method = rkf45, arrows = NONE);**

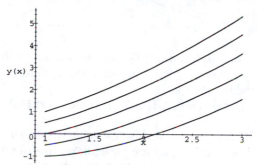

Numerical Solutions of Higher Order Differential Equations. In Chapter 7, we showed how to produce numerical solutions of first order initial value problems using **dsolve(. . ., numeric)**. Higher order initial value problems can be solved using essentially the same procedure. Here is an example using **dsolve(. . ., numeric)** to solve a third order initial value problem.

> **eqn2 := diff(y(x), x$3) = cos(y(x)) + 2*x^2;**

$$eqn2 := \frac{\partial^3}{\partial x^3} y(x) = cos(y(x)) + 2x^2$$

> **ini2 := y(0) = 1, D(y)(0) = 0, (D@@2)(y)(0) = 1;**

$$ini2 := y(0) = 1, D(y)(0) = 0, D^{(2)}(y)(0) = 1$$

> **sol2 := dsolve({eqn2, ini2}, y(x), numeric, maxfun = 2000, startinit = TRUE):**
> **evalf(sol2(1), 4);**

$$[x = 1.0, y(x) = 1.615, \frac{d}{dx} y(x) = 1.395, \frac{d^2}{dx^2} y(x) = 2.030]$$

Notice that evaluating the numerical solution produced by **dsolve(..., numeric)** yields a list containing the value of the numerical solution, as well as its first and second derivatives. We can use the method of Chapter 7 to plot the solution.

> **ya2 := u −> subs(sol2(u), y(x)):**
> **plot(ya2, 0..2);**

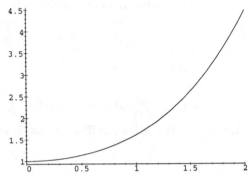

We could produce the same figure using **DEplot**, which also allows us to specify initial conditions for the derivatives in a higher order differential equation. The corresponding command is

> **DEplot(eqn2, y(x), x = 0..2, {[ini2]}, method = rkf45);**

More about DEplot.   In this section, we describe some ways to improve the graphs produced by **DEplot**. These techniques are particularly useful when dealing with solutions that are highly oscillatory or that have singularities.

Consider the differential equation

> **eqn3 := diff(x(t), t$2) = −x(t);**

$$eqn3 := \frac{\partial^2}{\partial t^2} x(t) = -x(t)$$

We know that a fundamental set of solutions of this equation is $\{\cos t, \sin t\}$. Here is what **DEplot** does with this equation.

> **DEplot(eqn3, x(t), 0..40, {[x(0) = 1, D(x)(0) = 4], method = rkf45});**

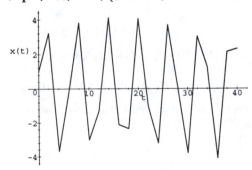

Notice how jagged the graph is. This is an indication that the step size used by **DEplot** is too large to capture the detailed behavior of the solution. The default step size used by *Maple* is the length of the specified interval divided by 20. Thus in the preceding example, the step size is $40/20 = 2$. We specify the plotting step size manually with the option **stepsize**. (In general, the plotting step size is not the step size used by the numerical method.) For example,

> **DEplot(eqn3, x(t), 0..40, {[x(0) = 1, D(x)(0) = 4]},**
> **method = rkf45, stepsize = 0.2);**

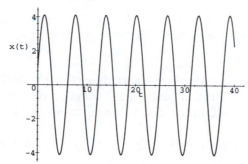

Much better!

Now consider the first order equation $y' = y^2$. The solution of this equation with initial condition $y(0) = y_0$ is

$$y(x) = -\frac{1}{x - \frac{1}{y_0}}.$$

Thus every solution (except the trivial one $y(x) = 0$), has a vertical asymptote at $x = 1/y_0$. If we try to plot a solution of this equation using **DEplot**, we get an error message.

> **eqn4 := diff(y(x), x) = y(x)^2;**

$$eqn4 := \frac{d}{dx}y(x) = (y(x))^2$$

> **DEplot(eqn4, y(x), x = 0..2, {[y(0) = 1]}, method = rkf45,**
> **arrows = NONE);**

```
Error, (in DEtools/DEplot/drawlines) Stopping integration
due to, rkf45 is unable to achieve requested accuracy
```

In this example, we have used the initial condition $y(0) = 1$, so the solution has a singularity at $x = 1$. The numerical method is unable to track the solution near the singularity, and so the plotting routine simply gives up. If we replace the **rkf45** method with the default method, we get an unexpected result.

> **DEplot(eqn4, y(x), x = 0..2, {[y(0) = 1]}, arrows = NONE);**

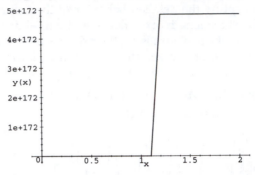

Using the default method, *Maple* has plotted past the singularity. Consequently, the values it shows are all ridiculously large. To get a correct picture, we must combine several options to **DEplot**. The following command works:

> **DEplot(eqn4, y(x), x = 0..2, {[y(0) = 1]}, y = 0..100, stepsize = 0.01,**

   method = rkf45, arrows = NONE);

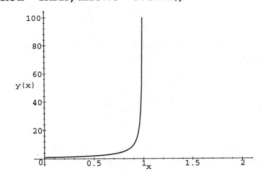

In order to get a reasonable plot using **rkf45**, we were forced both to bound the *y*-range on the plot, and to decrease the plotting step size. (Changing either option separately produces the same error message as above.) Basically, **DEplot** will keep marching toward the singularity until the solution jumps so much in a single step that it exceeds the desired bound on the error.

In practice, you might not know in advance if the solution of the differential equation has a singularity. An error message like the one above, or a plot where all the *y*-values are enormous, would tip you off to the existence of a singularity. You should then try to adjust the *y*-range and the step size to get a reasonable plot. Sometimes (for example, when trying to plot multiple solutions with singularities on the same graph) it is also necessary to limit the *x*-range. For an example, see the solution to Problem 3 of Problem Set C in the Sample Worksheet Solutions.

Troubleshooting

In this section, we collect some advice for avoiding mistakes that users of *Maple* commonly make. We also discuss the warning messages that *Maple* produces, and describe some techniques for recovering from errors.

The Most Common Mistakes. Here is a list of some of the most common mistakes in using *Maple*.

1. Failure to clear values. See the section *Clearing Values* above.
2. Missing semicolons at the end of input statements. *Maple* will not respond until you type a colon or semicolon.
3. Syntax errors, including mismatched braces, brackets, or parentheses. *Maple* usually catches these mistakes.
4. Failure to distinguish between objects such as functions, expressions, and equations. In particular, attempting to plot the wrong type of object.
5. Using a command from a *Maple* package without loading it. For example, typing **DEplot** without first typing **with(DEtools)**.
6. Typing the wrong kind of "=" sign. A plain "=" sign indicates an equation, while ":=" indicates an assignment, as in **x := 1**.
7. Improperly concatenating variables. For example, **xy** is a variable named xy and not x times y. The product of x and y must be written **x*y**.
8. Improper use of special functions. For example, to use the natural logarithm function in *Maple* you type **log(x)** and not **log** x. The exponential funtion is **exp(x)**.
9. Mistakes arising from the use of quotation marks (") to refer to the output of the previous command. The previous command is the one most recently executed, not necessarily the preceding command in the Worksheet.
10. Incorrectly defining a function using the syntax **f(x) := . . ..** See the section *Functions and Expressions* in this chapter.
11. Typing **y** instead of **y(x)** in **dsolve** and **dsolve(. . ., numeric)**.

Error and Warning Messages. Pay close attention to the error and warning messages that *Maple* generates. Although these messages may seem cryptic, they sometimes give valuable clues to locating mistakes. For example, suppose that we execute the following command without first defining a function $u(x)$.

```
> plot(u(x), x = 0..1);
Plotting error, empty plot
```

This strange error message is *Maple*'s way of telling you that you haven't really given it anything to plot. In this case, we tried to plot a function that hadn't been defined.

Occasionally, an error message tells you exactly what's wrong. For example,

> **eqn := diff(y(x), x) = y^2;**

$$eqn := \frac{d}{dx} y(x) = y^2$$

> **dsolve(eqn);**
```
Error, (in dsolve) dsolve uses a 2nd argument,vars,
which is missing
```

In this case, we've forgotten to specify the second argument. The correct syntax is **dsolve(eqn, y(x))**.

In many cases, error messages are due to errors made by the user. In other cases, these messages occur because *Maple* is being asked to do something that is either mathematically impossible, or exceeds the mathematical capabilities of the software. Here's a simple example.

> **1/0;**
```
Error, division by zero
```
Here's a more complicated example.

> **sol := dsolve({eqn, y(0) = 1}, y(x), numeric, maxfun = 4000);**

$$sol := proc(rkf45\_x)...end$$

> **sol(1);**
```
Error, (in sol) rkf45 is unable to achieve
requested accuracy
```

In this example, the solution of the initial value problem has a singularity at $x = 1$. When we enter **sol(1)**, we ask *Maple* to go beyond the singularity. It correctly refuses to do so, because it cannot maintain the desired degree of accuracy as it approaches the vertical asymptote.

Problem Set C

Numerical Solutions of Differential Equations

In this problem set you will use the *Maple* commands **dsolve(. . ., numeric)**, **plot**, and **DEplot** to solve numerically and plot solutions to ordinary differential equations. These commands are explained in Chapters 3, 7 and 8.

1. We are interested in the solution $y = \phi(x)$ to the initial value problem (IVP)

$$y' = x^2 + y^2, \qquad y(0) = 1. \qquad (i)$$

Observe that on the interval $0 \leq x \leq 1$, we have the inequalities

$$y^2 \leq x^2 + y^2 \leq 1 + y^2.$$

Therefore we must have $\phi_0(x) \leq \phi(x) \leq \phi_1(x)$, where ϕ_0, ϕ_1 are the solutions to the respective IVPs

$$\begin{cases} y' = y^2 \\ y(0) = 1, \end{cases} \qquad \begin{cases} y' = 1 + y^2 \\ y(0) = 1. \end{cases}$$

Solve these two IVPs explicitly and conclude that $\phi(x) \to \infty$ as $x \to x_*$, for some $x_* \in [\pi/4, 1]$. Then use *Maple* to compute a numerical solution of (i), and find an approximate value of x_* using **dsolve(. . ., numeric)** and **plot**.

2. We are interested in the solution $y = \phi(x)$ to the initial value problem (IVP)

$$y' = y^2 - x, \qquad y(0) = 2. \qquad (i)$$

Observe that on the interval $0 \leq x \leq 1$, we have the inequalities

$$y^2 \geq y^2 - x \geq y^2 - 1.$$

Therefore we must have $\phi_0(x) \geq \phi(x) \geq \phi_1(x)$, where ϕ_0, ϕ_1 are the solutions to the respective IVPs

$$\begin{cases} y' = y^2 \\ y(0) = 2, \end{cases} \qquad \begin{cases} y' = y^2 - 1 \\ y(0) = 2. \end{cases}$$

Solve these two IVPs explicitly and conclude that $\phi(x) \to \infty$ as $x \to x_*$, for some $x_* \in [0.5, 0.5 \ln 3]$. Then compute a numerical solution of (i) and find an approximate value of x_* using **dsolve(. . ., numeric)** and **plot**.

3. We shall study solutions $y = \phi_b(x)$ to the IVP

$$y' = (y - x)(1 - y^3), \qquad y(0) = b$$

for nonnegative values of x.

(a) Plot numerical solutions $\phi_b(x)$ for several values of b. Make sure to include values of b that are less than or equal to 0, between 0 and 1, equal to 1, and greater than 1.

(b) Now, on the basis of these plots, describe the behavior of the solution curves $\phi_b(x)$ for positive x, when $b \leq 0, 0 < b < 1, b = 1$, and $b > 1$. Identify limiting behavior and indicate where the solutions are increasing or decreasing.

(c) Finally, by combining your plots with a plot of the line $y = x$, show that the solution curves for $b > 1$ are asymptotic to this line. Explain from the differential equation why that is plausible. Plot the direction field of the differential equation to confirm your analysis.

4. We shall study solutions $y = \phi_b(x)$ to the IVP

$$y' = (y - x^2)(1 - y^2), \qquad y(0) = b$$

for nonnegative values of x.

(a) Plot numerical solutions $\phi_b(x)$ for several values of b. Make sure to include values of b that are less than -1, equal to -1, between -1 and 1, equal to 1, and greater than 1.

(b) Now, based on these plots, describe the behavior of the solution curves $\phi_b(x)$ for positive x, when $b < -1$, $b = -1$, $-1 < b < 1$, $b = 1$, and $b > 1$. Identify limiting behavior and indicate where the solutions are increasing or decreasing.

(c) Finally, by combining your plots with a plot of the parabola $y = x^2$, show that the solution curves for $b > 1$ are asymptotic to this parabola. Explain from the differential equation why that is plausible. Plot the direction field of the differential equation to confirm your analysis.

5. In this problem, we analyze the Gompertz-threshold model from Problem 13 of Problem Set B. That is, consider the differential equation

$$y' = y(1 - \ln y)(y - 3).$$

Using various nonnegative values of $y(0)$, find and plot several numerical solutions on the interval $0 \leq t \leq 6$. (Changing the range of the plot may be helpful here.) By examining the differential equation and analyzing your plots, identify all equilibrium solutions and discuss their stability.

6. Consider the IVP

$$\frac{dy}{dx} = \frac{x - e^{-x}}{y + e^y}, \qquad y(1.5) = 0.5.$$

(a) Use **dsolve(..., numeric)** to find approximate values of the solution at $x = 0, 1, 1.8$, and 2.1. Then plot the solution.

(b) If you did Problem 4 in Problem Set B, enter those results into this worksheet. Compare the values of the actual solution and the numerical solution at the four specified points. Plot the actual solution and the numerical solution on the same graph.

(c) Now plot the numerical solution on several large intervals (e.g., $0 \le x \le 10$ or $0 \le x \le 100$). Make a guess about the nature of the solution as $x \to \infty$. Try to justify your guess on the basis of the differential equation.

7. Consider the IVP

$$ e^x \sin y - 2y \sin x + (e^x \cos y + 2 \cos x) \frac{dy}{dx} = 0, \qquad y(0) = 0.5. $$

(a) Use **dsolve(..., numeric)** to find approximate values of the solution at $x = -1, 1$, and 2. Then plot the solution.

(b) If you did Problem 5 in Problem Set B, enter those results into this worksheet. Compare the values of the actual solution and the numerical solution at the three specified points. Plot the actual solution and the numerical solution on the same graph.

(c) Now plot the numerical solution on several large intervals (e.g., $0 \le x \le 10$ or $0 \le x \le 100$). Make a guess about the nature of the solution as $x \to \infty$. Try to justify your guess on the basis of the differential equation.

8. We know that the differential equation

$$ y' = e^{-x^2} $$

cannot be solved in terms of elementary functions.

(a) Use **dsolve** to solve this equation.

(b) Note the occurrence of the built-in function **erf**. Mathematics texts refer to this function as the *error function*. Use the *Maple* differentiation operator **diff** to see that

$$ \frac{d}{dx}(\text{erf}(x)) = \frac{2}{\sqrt{\pi}} e^{-x^2}. $$

In fact,

$$ \text{erf}(x) = \frac{2}{\sqrt{\pi}} \int_0^x e^{-t^2} \, dt. $$

(c) Although one does not have elementary formulas for this function, the numerical capabilities of *Maple* mean that we "know" this function as well as we "know" elementary functions like $\tan x$. To illustrate this, evaluate $\text{erf}(x)$ at $x = 0, 1$, and 10.5, and plot $\text{erf}(x)$ on $-10 \le x \le 10$.

(d) Compute $\lim_{x \to \infty} \operatorname{erf}(x)$ and $\int_{-\infty}^{\infty} e^{-t^2}\, dt$.

(e) Next solve the IVP

$$y' = 1 - 2xy, \qquad y(0) = 0$$

using the *Maple* command **dsolve**. What does the solution do for large x? Find the value x at which the solution takes its maximum. What is the maximum value?

(f) Compute $\partial f / \partial y$ for the function $f(x, y) = 1 - 2xy$. Discuss stability of the equation in (e). Plot solutions on the intervals $[-3, 0]$ and $[0, 3]$, using several different initial conditions at the left endpoints of each interval. Explain how these plots illustrate your conclusions about stability.

9. Consider the IVP

$$xy' + (\sin x)y = 0, \qquad y(0) = 1.$$

(a) Use **dsolve** to solve the IVP.

(b) Note the occurrence of the built-in function **Si**, called the *Sine Integral* function. Check that this function is an antiderivative of $\sin x / x$ by differentiating it. Evaluate $\lim_{x \to \infty} \operatorname{Si}(x)$.

(c) Plot $\operatorname{Si}(x)$. Discuss the features of the graph.

(d) Do the same with the solution to the IVP.

(e) Now solve the IVP using **dsolve(..., numeric)**, and plot the computed solution using **plot**.

You will find that *Maple* fails to deal with $\sin x / x$ at $x = 0$, even though the singularity is removable. One way to get around this is to give the initial condition $y = 1$ at a value of x extremely close to, but not equal to, zero. Since we are only finding an approximate solution with **dsolve(..., numeric)** anyhow, there shouldn't be much harm done as long as the amount by which we move the initial condition is small compared with the error we expect from the numerical procedure.

(f) Discuss the stability of the differential equation. Illustrate your conclusions by graphing solutions with different initial values on the interval $[-10, 10]$.

10. Solve the following equations numerically, then plot the solutions. On the basis of your plots, predict what happens to each solution as x increases. In particular, if there is a limiting value for y, either finite or infinite, then find it. If it is unclear from the plot you've made, try replotting on a larger interval. Another possibility is that the solution blows up in finite time. If so, estimate the time. If your solution curves appear jagged, use an option to adjust the number of points being plotted (**numpoints** with **plot** and **stepsize** with **DEplot**).

(a) $y' = e^{-2x} + \dfrac{1}{1 + y^2}$, $y(0) = -1$

(b) $y' = e^{-2x} + y^2$, $\quad y(0) = -1$

(c) $y' = \cos x - y^3$, $\quad y(0) = 0$

(d) $y' = (\sin x)y - y^2$, $\quad y(0) = 1$.

11. Solve the following equations numerically, then plot the solutions. On the basis of your plots, predict what happens to each solution as x increases. See Problem 10 for more instructions.

(a) $y' = 5x - 3\sqrt{y}$, $\quad y(0) = 2$

(b) $y' = (x^2 - y^2)\sin y$, $\quad y(0) = -1$

(c) $y' = \dfrac{y^2 + 2xy}{x^2 + 3}$, $\quad y(1) = 2$

(d) $y' = -2x + e^{-xy}$, $\quad y(0) = 1$.

12. This problem illustrates one of the possible pitfalls of blindly applying numerical methods without paying attention to the theoretical aspects of the differential equation itself. Consider the equation

$$-4x^2 + 2y(x) + x\,y'(x) = 0.$$

(a) Use the *Maple* program in Chapter 7 to compute the Euler Method approximation to the solution with initial condition $y(-0.5) = 4.25$, using step size $h = 0.2$ and $n = 10$ steps. The program will generate a list of ordered pairs of numbers representing the (x, y) coordinates of points of the approximate solution. Use **plot** to obtain a piecewise linear graph of the approximate solution.

(b) Now modify the program to implement the Improved Euler Method. Repeat part (a) using the Improved Euler Method. Can you make sense of your answers?

(c) Next try to use **dsolve(. . ., numeric)** to find an approximate solution on the interval $(-0.5, 1.5)$, and plot it with **plot**. What is the interval on which the approximate solution is defined?

(d) Solve the equation explicitly and graph the solutions for the initial conditions $y(0) = 0$, $y(-0.5) = 4.25$, $y(0.5) = 4.25$, $y(-0.5) = -3.75$, and $y(0.5) = -3.75$. Now explain your results in (a)–(c). Could we have known, without solving the equation, whether to expect meaningful results in parts (a) and (b)? Why? Can you explain how **dsolve(. . ., numeric)** avoids making the same "mistake"?

13. Consider the IVP

$$\frac{dy}{dx} = e^{-x} - 2y, \qquad y(-1) = 0.$$

(a) Use the *Maple* program in Chapter 7 to compute the Euler Method approximation to $y(x)$ with step size $h = 0.5$ and $n = 4$ steps. The program will generate a list of ordered pairs (x_i, y_i). Use **plot** to display the piecewise linear function connecting the points (x_i, y_i). Repeat with $h = 0.2$ and $n = 10$.

(b) Now modify the program to implement the Improved Euler Method, and repeat part (a) using the Improved Euler Method.

(c) Find the exact solution of the IVP and plot the solution, the two Euler approximations, and the two Improved Euler approximations on the same graph. Label the five curves.

(d) Now use the Euler Method program with $h = 0.5$ to approximate the solution on the interval $[-1, 9]$. Plot both the approximate and exact solutions on this interval. How close is the approximation to the exact solution as x increases? In light of the discussion of stability in Chapters 5 and 7, explain your results in parts (a)–(c).

14. Consider the IVP

$$\frac{dy}{dx} = y - 4e^{-3x}, \qquad y(0) = 1.$$

(a) Use the *Maple* program in Chapter 7 to compute the Euler Method approximation to $y(x)$ with step size $h = 0.5$ and $n = 6$ steps. The program will generate a list of ordered pairs (x_i, y_i). Use **plot** to display the piecewise linear function connecting the points (x_i, y_i). What appears to be happening to y as x increases?

(b) Repeat part (a) with $h = 0.2$ and $n = 15$, and then with $h = 0.1$ and $n = 30$. How are the approximate solutions changing as the step size decreases? Can you make a reliable prediction about the long-term behavior of the solution?

(c) Use **dsolve(..., numeric)** to find an approximate solution and plot it on the interval $[0, 3]$. Now what does it look like y is doing as x increases? Next plot the solution from **dsolve(..., numeric)** on a larger interval (going at least to $x = 20$). Again, what is happening to y as x increases?

(d) Solve the IVP explicitly and compare the exact solution to the approximations found above. In light of the discussion of stability in Chapters 5 and 7, explain your results in parts (a)–(c).

15. Consider the IVP

$$\frac{dy}{dx} = 2y - 2 + 3e^{-x}, \qquad y(0) = 0.$$

(a) Use the *Maple* program in Chapter 7 to compute the Euler Method approximation to $y(x)$ with step size $h = 0.2$ and $n = 10$ steps. The program will generate a list of ordered pairs (x_i, y_i). Use **plot** to display the piecewise linear function connecting the points (x_i, y_i) What appears to be happening to y as x increases?

(b) Repeat part (a) with $h = 0.1$ and $n = 20$, and then with $h = 0.05$ and $n = 40$. How are the approximate solutions changing as the step size decreases? Can you make a reliable prediction about the long-term behavior of the solution?

(c) Use **dsolve(..., numeric)** to find an approximate solution and plot it on the interval $[0, 2]$. Now what does it look like y is doing as x increases? Next plot the solution from **dsolve(..., numeric)** on a larger interval (going at least to $x = 10$). Again, what is happening to y as x increases?

(d) Solve the IVP explicitly and compare the exact solution to the approximations found above. In light of the discussion of stability in Chapters 5 and 7, explain your results in parts (a)–(c).

16. We know that e^x is the solution to the IVP

$$\frac{dy}{dx} = y, \qquad y(0) = 1,$$

so if we solve this IVP numerically we get approximate values for e^x. Use **dsolve(..., numeric)**, employing the accuracy option discussed in Chapter 7, to calculate values for $e^{0.1}, e^{0.2}, \ldots, e$ that have 15 correct digits. Present your results in a table. In a second column print the values of e^x for $x = 0.1, 0.2, \ldots, 1$, obtained by using the built-in function **exp**. Show at least 15 digits. Compare the two columns of values.

Chapter 9

Qualitative Theory of Second Order Linear Equations

Newton's second law of dynamics—force is equal to mass times acceleration—tells physicists that, in order to understand how the world works, they must pay attention to the forces. Since acceleration is a second derivative, the law also tells us that second order differential equations are likely to appear when we apply mathematics to study the real world.

The simplest second order differential equations are linear equations with *constant coefficients*:

$$ay'' + by' + cy = g(x).$$

These equations model a wide variety of physical situations, including oscillating springs, simple electric circuits, and the vibrations of tuning forks to produce sound and of electrons to produce light. In other situations, such as the motion of a pendulum, we may be able to approximate the resulting differential equation reasonably well by a linear differential equation with constant coefficients. Fortunately, we know (and *Maple* can apply) several techniques for finding explicit solution formulas to linear differential equations with constant coefficients.

Unfortunately, we often cannot find solution formulas for more general second order equations. We do not even have good methods for finding solution formulas for second order linear differential equations with *variable coefficients*,

$$y'' + p(x)y' + q(x)y = g(x),$$

which also have important applications to physics. For example, Airy's equation,

$$y'' - xy = 0, \tag{1}$$

arises in diffraction problems in optics, and Bessel's equation,

$$y'' + \frac{1}{x}y' + y = 0, \tag{2}$$

occurs in the study of vibrations of a circular membrane and of water waves with circular symmetry.

When studying first order differential equations for which exact solution formulas were unavailable, we could turn to several other methods. By specifying an initial value, we could compute an approximate numerical solution. By letting the initial values vary, we could plot a one-parameter family of approximate solution curves and get a feel for the behavior of a general solution. Alternatively, we could plot the direction field of the differential equation and use it to draw conclusions about the qualitative behavior of solutions.

For second order equations, these methods are more cumbersome; you must specify two conditions to pick out a unique solution. Most commonly, we specify initial values for both the function and its first derivative at some point. Then we can use numerical techniques to compute an approximate solution. Now in order to graph enough solutions to get a good idea of the general behavior of solution curves, we must construct a two-parameter family of solutions. In some applications, however, the differential equation comes with *boundary conditions; i.e.,* we specify values $y(x_0) = y_0$ and $y(x_1) = y_1$ at two distinct points. The resulting problem is called a boundary value problem. In this situation, we cannot directly compute a numerical solution; we know the value, but not the slope with which to start. For the same reason—the initial value of the solution does not determine the initial slope—we cannot draw a direction field for a second order equation.

In this chapter, we describe how to use *Maple* to find exact solutions of second order linear differential equations with constant coefficients. We also describe how to find and plot numerical solutions to more general second order differential equations. In addition, we describe a method for solving boundary value problems using *Maple*'s numerical solver. Finally, we introduce comparison methods and a more sophisticated geometric method for analyzing second order linear equations with variable coefficients. These qualitative methods have an advantage over the more obvious numerical and graphical methods. They more effectively yield information about properties shared by all solutions of a differential equation, about the oscillatory nature of solutions of an equation, or about the precise rate of decay or growth of solutions.

Comparison methods provide information on the solutions of a variable coefficient equation by comparing the equation with an appropriate constant coefficient equation. This possibility is suggested by the following example. For large x, the coefficient $1/x$ in Bessel's equation (2) is close to 0. So, Bessel's equation is close to the equation $y'' + y = 0$, whose general solution can be written as $y = R\cos(x - \delta)$. Thus, one might expect solutions to Bessel's equation to oscillate and look roughly like sine waves for large x. We present a result that validates such comparisons.

The geometric method is based on direction fields, which are not directly applicable to a second order equation. We show, however, how to construct a related first order equation whose direction field yields information about the solutions of the second order equation.

Second Order Equations with *Maple*

Maple's usual tools for finding symbolic solutions of differential equations work perfectly well for second order linear differential equations with constant coefficients. The syntax of the commmands is the same, except that we must specify more initial conditions.

EXAMPLE 1. Consider the differential equation

$$y'' + y' - 6y = 20e^x.$$

If we type

ode1 := diff(y(x), x$2) + diff(y(x), x) − 6*y(x) = 20*exp(x):
dsolve(ode1, y(x));

we get the solution

$$y(x) = -5e^x + \_C1e^{(-3x)} + \_C2e^{(2x)}.$$

As expected, the general solution depends on two arbitrary constants. To solve the differential equation with initial conditions $y(0) = 0$ and $y'(0) = 1$, we would type

dsolve({ode1, y(0) = 0, D(y)(0) = 1}, y(x));

The specific solution with these initial values is

$$y(x) = -e^x + \frac{4}{5}e^{(-3x)} + \frac{21}{5}e^{(2x)}.$$

    Maple can also solve certain boundary value problems. Let's continue using the same differential equation for our example, but take the boundary conditions $y(0) = 0$ at the first point and $y(\ln(2)) = 10$ at the second point. If we type

dsolve({ode1, y(0) = 0, y(ln(2)) = 10}, y(x));

we get the answer

$$y(x) = -5e^x + 5e^{(2x)}.$$

EXAMPLE 2. Now consider the differential equation

$$y'' + x^2 y' + y = 0,$$

which is a second order linear differential equation with variable coefficients. *Maple* is unable to find an explicit formula for the general solution. The most straightforward method for understanding the behavior of a general solution of the equation is to plot numerical solutions corresponding to a wide range of initial values and initial slopes. Here are the *Maple* commands to do this.

ode2 := diff(y(x), x$2) + x^2*diff(y(x), x) + y(x) = 0:
iniset := {seq(seq([y(0) = y0, D(y)(0) = yp0/2], y0 = −2..2), yp0 = −2..2):
with(DEtools):
DEplot(ode2, y(x), −2..3, iniset, method = rkf45);

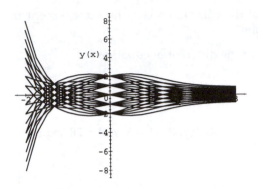

Figure 1

The result is shown in Figure 1. We see that the solutions seem to decay for positive x, and seem to blow up for negative x.

Could we have obtained this qualitative information directly from the differential equation, instead of from the graph of a few solutions? One approach is to look at the solutions of differential equations that are similar to this one. For example, when x is small, the x^2 term is very close to zero. So the solutions might be close to the solutions of the differential equation

$$y'' + y = 0.$$

We know that the solutions to this equation are sine curves. For example, the solution satisfying the initial conditions $y(0) = 0$, $y'(0) = 1$ is just $\sin(x)$. So, one might expect the solutions of this initial value problem to look like $\sin(x)$ for small values of x. In Figure 2, we have plotted $\sin(x)$ and the numerical solution to the initial value problem on the same axes.

The graph of the numerical solution in Figure 2 appears to level off as x increases. Looking again at the differential equation, we see that when x is large, the $x^2 y'$ term should dominate the y term. So, the solutions to the differential equation should be close to those of

$$y'' + x^2 y' = 0.$$

This equation can almost be solved explicitly; its general solution is given by

$$y(x) = A + B \int_0^x e^{-u^3/3} \, du.$$

When x is large and positive, the exponential function being integrated is nearly zero, so the solutions should become almost constant. When x is large and negative, the exponential term is large, and we expect the solutions to blow up as well.

Finally, let's look at the boundary value problem

$$y'' + x^2 y' + y = 0, \quad y(0) = 0, \quad y(1) = 1.$$

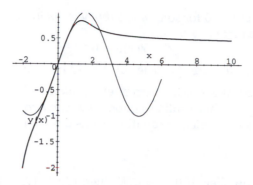

Figure 2

Here we have the same differential equation, but we have specified values for the solution curve at two different points. Because it is a boundary value problem rather than an initial value problem, we cannot just invoke the numerical solver directly. Instead, consider the following command, which tells *Maple* to compute a numerical solution as a function of the first derivative:

> nsol2 := s –> dsolve({ode2, y(0) = 0, D(y)(0) = s}, y(x), numeric,
>   maxfun = 1000, startinit = TRUE):
> trial := s –> nsol2(s)(1);

Now when we evaluate **nsol2** at some real number, we get a procedure that can be used to compute numerical solutions to the differential equation. When we evaluate that procedure at $x = 1$, as is done by the **trial** command, we find the approximate value of that numerical solution at the desired endpoint in our initial value problem. After some trial and error, one finds in this case that **trial(1.278)** produces the output

$$[x = 1, y(x) = .9998391792710923, \frac{\partial}{\partial x} y(x) = .4495616548265483].$$

So, the solution to the boundary value problem is (approximately) the numerical solution with initial conditions $y(0) = 0$ and $y'(0) = 1.278$.

Comparison Methods

In this section, we discuss the Sturm Comparison Theorem. This theorem is a precise form of the rough comparisons we carried out in the previous section. In addition, we will discuss the relation between the zeros of two linearly independent solutions of a second order linear equation.

STURM COMPARISON THEOREM. *Let $u(x)$ be a solution to the equation*

$$y'' + q(x)y = 0,$$

and suppose $u(a) = u(b) = 0$ *for some* $a < b$ *(but u is not identically zero). Let* $v(x)$ *be a solution to the equation*

$$y'' + r(x)y = 0,$$

where $r(x) \geq q(x)$ *for all* $a \leq x \leq b$. *Then* $v(x) = 0$ *for some* x *in* $[a, b]$.

We will defer the proof of this theorem briefly, preferring instead to explain how to use it. We typically use this result to compare a variable coefficient equation with an appropriate constant coefficient equation. Consider, for example, the equation

$$y'' + \frac{1}{x}y = 0. \tag{3}$$

Let K be a positive number. If $0 < x \leq K$, then $1/x \geq 1/K$. We now apply the above theorem with $q(x) = 1/K$ and $r(x) = 1/x$. Thus, we compare equation (3) to the constant coefficient equation $y'' + (1/K)y = 0$, whose general solution is $u(x) = R\cos(\sqrt{1/K}x - \delta)$. Given any interval in $(0, K]$ of length $\pi\sqrt{K}$, the value of δ can be chosen so that $u(x) = 0$ at the endpoints of the interval. Then the Sturm Comparison Theorem implies that every solution of (3) has a zero on this interval. This argument implies that on $(0, K]$, the zeros of (3) cannot be farther apart than $\pi\sqrt{K}$.

In particular, solutions of (3) must oscillate as $x \to \infty$, though the oscillations may become less frequent as x increases. Indeed, by turning the above comparison around, one concludes that the zeros of solutions of (3) in $[K, \infty)$ must be at least $\pi\sqrt{K}$ apart.

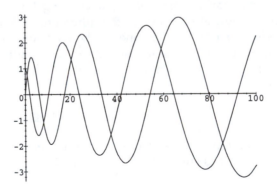

Figure 3

Figure 3 shows two representative solutions of (3) computed with **dsolve**. As you can see, the predictions of the previous paragraph are confirmed by the graph.

The Sturm Comparison Theorem can also be used to study the zeros of Airy's equation and of Bessel's equation. In the latter case, a substitution is made to eliminate the y' term. See Problems 15 and 16 in Problem Set D.

The Interlacing of Zeros. You may have noticed that the curves in Figure 3 take turns crossing the horizontal axis. We say that the zeros of the two solutions are *interlaced*: between any two zeros of one solution there is a zero of the other. This is a common phenomenon—the property of interlaced zeros holds for any pair of linearly independent solutions of a homogeneous second order linear differential equation

$$y'' + p(x)y' + q(x)y = 0. \tag{4}$$

If $p(x) = 0$, as is the case in equation (1), the proof of the interlacing result follows from an application of the Sturm Comparison Theorem by comparing the equation with itself. For the more general equation (4), the proof is based on the Wronskian. Let $y_1(x)$ and $y_2(x)$ be any two linearly independent solutions of (4). Recall that the Wronskian of $y_1(x)$ and $y_2(x)$ is

$$W(x) = y_1(x)y_2'(x) - y_1'(x)y_2(x).$$

Since y_1 and y_2 are solutions of (4), it follows that W satisfies the differential equation

$$W' = -p(x)W,$$

whose general solution is

$$W(x) = Ce^{-\int p(x)dx}.$$

This formula shows that $W(x)$ does not change sign—it always has the same sign as the constant C, which must be nonzero (or else y_1 and y_2 would be linearly dependent).

Assume $y_1(a) = y_1(b) = 0$, and that there are no zeros of y_1 between a and b. Then

$$W(a) = -y_1'(a)y_2(a),$$
$$W(b) = -y_1'(b)y_2(b).$$

Since y_1 does not change sign between a and b, we know $y_1'(a)$ and $y_1'(b)$ have opposite signs. Then, since $W(a)$ and $W(b)$ have the same sign, $y_2(a)$ and $y_2(b)$ must have opposite signs. Therefore y_2 must be zero somewhere between a and b. Similarly, y_1 must have a zero between any two zeros of y_2.

EXERCISE. The solutions to Airy's equation and Bessel's equation are considered so important that they have been given names in *Maple*. The Airy functions $Ai(x)$ and $Bi(x)$ are a fundamental set of solutions for (1), and the Bessel functions $J_0(x)$ and $Y_0(x)$ are a fundamental set of solutions for (2). These functions are built into *Maple* as **AiryAi(x)**, **AiryBi(x)**, **BesselJ(0, x)**, and **BesselY(0, x)**. Use *Maple* to check that the zeros of the two solutions **AiryAi(x)** and **AiryBi(x)** to Airy's equation are interlaced. Do the same for the Bessel functions **BesselJ(0, x)** and **BesselY(0, x)**.

Proof of the Sturm Comparison Theorem. The proof is based on the Wronskian, as was the proof of the interlacing result. Recall the hypotheses: $u'' + q(x)u = 0$ and $v'' + r(x)v = 0$, with $q(x) \leq r(x)$ between two zeros a and b of u. Since u is not identically zero, we can assume u does not change sign between a and b. (If it did it would have a zero between a and b, and we could look at a smaller interval.) Assume, for instance, that $u > 0$ between a and b. Since $u = 0$ at a and b, it follows that $u'(a) > 0$ and $u'(b) < 0$. (Notice that these derivatives cannot be zero because then by the uniqueness theorem u would be identically zero.) We want to prove that v is zero somewhere in $[a, b]$; we suppose it is not, say $v(x) > 0$ throughout $[a, b]$, and argue to obtain a contradiction.

Let $W(x)$ be the Wronskian of u and v,

$$W(x) = u(x)v'(x) - u'(x)v(x);$$

then

$$W(a) = -u'(a)v(a) < 0,$$
$$W(b) = -u'(b)v(b) > 0.$$

Also, since $q(x) \leq r(x)$,

$$W' = uv'' - u''v = u(-r(x)v) - (-q(x)u)v = (q(x) - r(x))uv \leq 0.$$

But this is impossible—the last inequality implies W does not increase between a and b, yet $W(a) < 0 < W(b)$. This contradiction means that v cannot have the same sign throughout $[a, b]$, and therefore v must be zero somewhere in the interval.

A Geometric Method

As mentioned above, direction fields are not directly applicable to second order equations. We now show, however, that with a given second order homogeneous equation, we can associate a first order equation whose direction field yields information about the solutions of the second order equation. (Our approach is similar to the classical method of associating a first order Riccati equation to a second order linear equation. The substitution we use is akin to the Prüfer substitution for Sturm-Liouville systems; see G. Birkhoff and G.-C. Rota, **Ordinary Differential Equations**, 3rd ed., J. Wiley and Sons, Inc., 1978.)

Consider the homogeneous equation

$$y'' + p(x)y' + q(x)y = 0. \tag{5}$$

Let

$$z = \arctan\left(\frac{y}{y'}\right);$$

then

$$z' = \left(1 + \left(\frac{y}{y'}\right)^2\right)^{-1} \frac{d}{dx}\left(\frac{y}{y'}\right) = \frac{y'^2}{y'^2 + y^2} \frac{y'^2 - yy''}{y'^2} = \frac{y'^2 - yy''}{y'^2 + y^2}.$$

Also, since $\tan z = y/y'$, notice that

$$\sin z = \frac{y}{\sqrt{y'^2 + y^2}}, \quad \cos z = \frac{y'}{\sqrt{y'^2 + y^2}}.$$

Then if y satisfies (5), and y is not identically zero, it follows that

$$z' = \frac{y'^2 - yy''}{y'^2 + y^2} = \frac{y'^2 - y(-p(x)y' - q(x)y)}{y'^2 + y^2} = \frac{y'^2 + p(x)yy' + q(x)y^2}{y'^2 + y^2}.$$

In other words,

$$z' = \cos^2 z + p(x)\sin z \cos z + q(x)\sin^2 z. \tag{6}$$

We have shown that for every solution y of (5) that is not identically zero, there is a corresponding solution z of (6). (This is not a one-to-one correspondence—every constant multiple of a solution y corresponds to the same solution z.) Although the solution curves of (6) will be different from the solution curves of (5), we now show that we can learn about the solutions of (5) by studying the solutions of (6); specifically, by considering the direction field of (6).

The Constant Coefficient Case. We begin by considering the equation

$$y'' - y = 0,$$

whose general solution is

$$y = c_1 e^x + c_2 e^{-x}.$$

The corresponding first order equation is

$$z' = \cos^2 z - \sin^2 z.$$

In Figure 4, we show the direction field of this equation, which we produced with the *Maple* command

```
with(DEtools):
dfieldplot(diff(z(x), x) = cos(z(x))^2 − sin(z(x))^2, z(x),
    x = 0..10, z = −Pi/2..Pi/2, arrows = LINE, axes = BOXED);
```

Notice that we plot z from $-\pi/2$ to $\pi/2$; this is the range of values taken on by the arctan function. You may wonder what happens if z goes off the bottom of the graph. The answer is that z "wraps around" to the top of the graph. Recall that $z = \arctan(y/y')$; the points $z = \pm\pi/2$ correspond to $y' = 0$, and when y' changes sign, then so does z by passing from $-\pi/2$ to $\pi/2$ (or vice versa).

Next, observe that there is an unstable equilibrium at $z = -\pi/4$, which corresponds to $y/y' = \tan(-\pi/4) = -1$; this represents solutions of the form $y = ce^{-x}$. Similarly, the stable equilibrium $z = \pi/4$ corresponds to $y/y' = \tan(\pi/4) = 1$, which represents the solutions $y = ce^x$. The fact that most solutions of the equation satisfied by z (those with initial condition other than $-\pi/4$) approach the stable equilibrium $z = \pi/4$ corresponds to the fact that most solutions $y = c_1 e^x + c_2 e^{-x}$ (those with $c_1 \neq 0$) grow like e^x as x increases.

Another basic example is

$$y'' + y = 0,$$

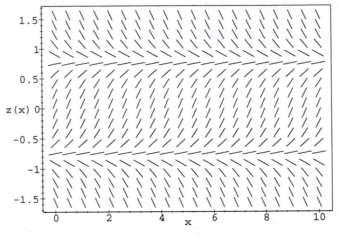

Figure 4

for which the corresponding first order equation is

$$z' = \cos^2 z + \sin^2 z = 1.$$

That is, z simply increases linearly, though every time it reaches $\pi/2$ it wraps around to $-\pi/2$. This corresponds to the oscillation of solutions y; for every time $z = \arctan(y/y')$ passes through zero, then so must y, and vice versa.

EXERCISE. Consider the general second order linear homogeneous equation with constant coefficients:

$$ay'' + by' + c = 0.$$

Investigate how the roots of the characteristic equation, if real, correspond to equilibrium solutions for z. Show that if the roots of the characteristic equation are complex, then z' is always positive.

The Variable Coefficient Case.   The examples above suggest ways to draw parallels between solutions z of (6) and solutions y of (5). First of all, if z is positive, then y and y' have the same sign. This implies that y is moving away from zero as x increases. We say in this case that y is *growing*, meaning that $|y|$ is increasing. Similarly, if z is negative then y and y' have opposite signs, which implies that y is moving toward zero as x increases—we say in this case that y is *decaying*, meaning that $|y|$ is decreasing. If z changes sign by passing through zero, then so does y, and if z passes through $\pm\pi/2$, then y' changes sign, showing that y has passed through a local maximum or minimum.

   In light of these observations, let us summarize what we can predict about the *long-term behavior* of a solution y of (5) in terms of the corresponding solution z of (6).

- If z remains positive as x increases, then y grows away from zero.
- If z remains negative as x increases, then y decays toward zero.
- If z continues to increase from $-\pi/2$ to $\pi/2$ and wraps around to $-\pi/2$ again, then y *oscillates* as x increases.

Notice from (6) that $z' = 1$ whenever $z = 0$, so it is not possible for z to decrease through zero. Thus the long-term behavior of z will usually fall into one of the three categories above.

We can be more precise about the growth or decay rate of y in cases where z approaches a limiting value. If the limiting value is θ, then y/y' approaches $\tan\theta$ as x increases. In other words, $y' \approx (\cot\theta)y$ for large x. Thus

$$y \approx c e^{(\cot\theta)x}$$

for large x, and $\cot\theta$ is the asymptotic exponential growth (or decay) rate of y. If z approaches zero as x increases, then y/y' approaches zero as well. One can show that y grows (if $z > 0$) or decays (if $z < 0$) faster than any exponential function. Similarly, if z approaches $\pi/2$, then y grows slower than exponentially, and if z approaches $-\pi/2$, then y decays slower than exponentially. Since y'/y approaches zero in these cases, y could grow or decay toward a finite, nonzero value.

Let us now see what the direction field of (6) tells us about solutions of Airy's and Bessel's equations.

Airy's Equation.   For Airy's equation,

$$y'' - xy = 0, \tag{7}$$

the corresponding first order equation is

$$z' = \cos^2 z - x\sin^2 z.$$

Figure 5 shows the direction field of this equation, plotted by a *Maple* command similar to the one used for Figure 4.

Notice that z is increasing steadily for negative x, so solutions of (7) must oscillate for negative x. For positive x, it is evident that solutions of (7) cannot oscillate; z passes through zero at most once, after which it remains positive. Hence the corresponding solution of (7) grows away from zero as x increases. It appears that once z becomes positive, it decreases to zero, indicating that most solutions of (7) grow faster than exponentially.

We now know there are solutions of (7) that grow very fast as x increases, and there are no oscillating solutions for positive x. Is it possible to have a decaying solution of (7)? Equivalently, is it possible for z to remain negative as x increases? Figure 5 strongly suggests that there is such a solution z, and hence that there is a solution y to (7) that decays toward zero at a rate faster than exponential.

The Airy function $\mathrm{Ai}(x)$ is, by definition, the unique (up to a constant multiple) solution of (7) that decays as x increases. As expected, the other Airy function

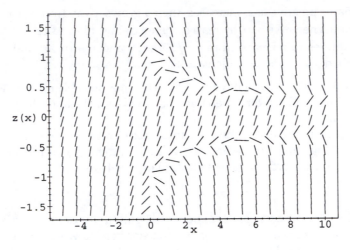

Figure 5

Bi(x) grows as x increases. You can see what these functions look like by plotting **AiryAi(x)** and **AiryBi(x)** in *Maple*. The general solution of (7) is a linear combination of Ai(x) and Bi(x).

Bessel's Equation. For Bessel's equation

$$y'' + \frac{1}{x}y' + y = 0, \tag{8}$$

the corresponding first order equation is

$$z' = \cos^2 z + \frac{1}{x}\sin z \cos z + \sin^2 z.$$

Figure 6 shows the direction field of this equation.

The picture is simpler than the one for Airy's equation. When x is away from zero, z is increasing steadily. So, solutions of (8) oscillate for both positive and negative x. Near $x = 0$, the direction field is irregular because of the $1/x$ term in the differential equation. We can't tell what happens from this picture, but there is reason to expect solutions of (8) will not behave nicely near this singularity in the differential equation.

One question this approach cannot answer easily is whether the amplitude of oscillating solutions grows, decays, or remains steady as x increases. Since the coefficient of the y' term in (8) is positive, we can expect solutions of (8) to behave like a physical oscillator with damping—the amplitude of oscillations should decrease as x increases. Since the damping coefficient goes to zero as x increases, we cannot be sure (without a more refined analysis) whether the

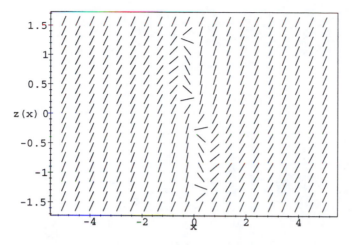

Figure 6

amplitude of solutions of (8) must decrease to zero. In fact it does, as you can check by plotting the Bessel functions **BesselJ(0, x)** and **BesselY(0, x)** in *Maple*.

Other Equations. Here are some other differential equations for which you can try this method.

- Bessel's equation of order n,

$$y'' + \frac{1}{x}y' + \left(1 - \frac{n^2}{x^2}\right)y = 0,$$

  for various n (above we studied the case $n = 0$). You can check your predictions using the *Maple* functions **BesselJ(n, x)** and **BesselY(n, x)**. These functions are considered in Problem 18 of Problem Set D.

- The modified Bessel equation of order n,

$$y'' + \frac{1}{x}y' - \left(1 + \frac{n^2}{x^2}\right)y = 0,$$

  for various n. The *Maple* functions **BesselI(n, x)** and **BesselK(n, x)** are a fundamental set of solutions for this equation.

- The parabolic cylinder equation

$$y'' + \left(n + \frac{1}{2} - \frac{1}{4}x^2\right)y = 0$$

  for various n. This equation arises in quantum mechanics, but its solutions are not built into *Maple*; you can check your predictions against approximate solutions from **dsolve(..., numeric)** instead. This equation is considered in Problem 17 of Problem Set D.

Problem Set D

Second Order Equations

1. Airy's equation is the linear second order homogeneous equation $y'' = xy$. Although it arises in a number of applications, including quantum mechanics, optics and waves, it cannot be solved exactly by the standard symbolic methods. In order to analyze the solution curves, let us reason as follows.

   (a) For x close to zero, the equation resembles $y'' = 0$, which has general solution $y = c_1 x + c_2$. We refer to this as a "facsimile" solution. Using **DEplot**, graph a numerical solution to Airy's equation with initial conditions $y(0) = 0$, $y'(0) = 1$, and the facsimile solution (with the same initial data) on the interval $(-2, 2)$. How well do they match?

   (b) For $x \approx -K^2 << 0$, the equation resembles $y'' = -K^2 y$, and the corresponding facsimile solution is given by $y = c_1 \sin(Kx + c_2)$. Again using the initial conditions $y(0) = 0$, $y'(0) = 1$, use **DEplot** to plot a numerical solution of Airy's equation over the interval $(-18, -14)$. Using the value $K = 4$, try to find values of c_1 and c_2 so that the facsimile solution matches well with the actual solution. Why shouldn't we expect the initial conditions for Airy's equation to be the appropriate initial conditions for the facsimile solution?

   (c) For $x \approx K^2 >> 0$, Airy's equation resembles $y'' = K^2 y$, which has solution $y = c_1 \sinh(Kx + c_2)$. (The hyperbolic sine function is called **sinh(x)** in *Maple*.) Plot a numerical solution of Airy's equation together with a facsimile solution (with $K = 4$) on the interval $(14, 18)$. In analogy with part (b), you have to choose values for c_1 and c_2 in the facsimile solution. (Try $c_1 = 1$, $c_2 = 0$.)

   (d) Plot the numerical solution of Airy's equation on the interval $(-20, 2)$. What does the graph suggest about the frequency and amplitude of oscillations as $x \to -\infty$? Could any of that information have been predicted from the facsimile analysis?

2. Consider Bessel's equation of order zero

$$x^2 y'' + x y' + x^2 y = 0 \qquad (i)$$

with initial data $y(0) = 1$, $y'(0) = 0$. The solution $J_0(x)$ is usually called the Bessel function of order zero of the first kind. Strictly speaking, this equation has a singularity at $x = 0$. However, this is one instance of a solution to a linear equation that exists outside the expected domain of definition. Still, as you will see below, the presence of the singularity can affect your computations.

(a) For x close to 0, x^2 is very small compared to x. The equation (i) is therefore approximately $xy' = 0$. Solve this equation with the preceding initial data. What does this "facsimile" solution to the original problem suggest to you about the behavior of J_0 near the origin?

(b) For x large and positive, x is small compared to x^2; and so we may approximate (i) by the equation $x^2(y'' + y) = 0$. Solve this equation with the same initial data. What does this suggest about the nature of the function J_0 for large x?

(c) Still thinking of x as large and positive, rewrite (i) in the form

$$y'' + \frac{1}{x}y' + y = 0.$$

If $x \approx K \gg 0$, we might approximate the equation by the constant coefficient equation

$$y'' + \frac{1}{K}y' + y = 0.$$

Solve that equation and see what further information you obtain about J_0. (If need be, choose a specific value for K, say $K = 100$, and examine the solution in order to refine your conclusion from (b).) Now explain why J_0 must be an even function, and so make the same deduction for large negative x.

(d) Throw caution to the winds and try **dsolve** on the original problem. You may be surprised to learn that *Maple* can handle it. Graph the solution function and see how well it manifests the qualitative properties you inferred in parts (a)–(c).

(e) Have a crack at the IVP with **dsolve(. . ., numeric)**. Can you find the value of the solution at $x = 1$? Explain your results.

3. This and some of the following problems concern models for the motion of a pendulum, which consists of a weight attached to a rigid arm of length L that is free to pivot in a complete circle. Neglecting friction and air resistance, we see that the angle $\theta(t)$ that the arm makes with the vertical direction satisfies the differential equation

$$\theta''(t) + \frac{g}{L}\sin(\theta(t)) = 0, \qquad (i)$$

where $g = 32.2$ ft/sec^2 is the gravitational acceleration constant. We will assume the arm has length 32.2 ft and so replace (i) by the simpler form

$$\theta'' + \sin\theta = 0. \qquad (ii)$$

(Alternatively, one can rescale time, replacing t by $\sqrt{g/L}\,t$, to convert (i) to (ii).) For motions with small displacements (θ small), $\sin\theta \approx \theta$, and ($ii$) can be approximated by the linear equation

$$\theta'' + \theta = 0. \tag{iii}$$

This equation has general solution $\theta(t) = A\cos(t - \delta)$, with *amplitude A* and *phase shift δ*. Since all the solutions to the linear approximation (iii) have period 2π, we expect that, for small displacements, the solutions to the true pendulum (ii) should have period close to 2π. One can investigate how the period depends on the amplitude A by plotting a numerical solution of equation (ii) using initial conditions $\theta(0) = A, \theta'(0) = 0$ on an appropriate interval for various A. Estimate, to within 0.1, the period of the pendulum for the amplitudes $A = 0.1$, $0.7, 1.5,$ and 3.0. What is happening to the accuracy of the linear approximation as the initial displacement increases?

4. In this problem, we'll look at what the pendulum does for various initial velocities (*cf.* Problem 3). Define a *Maple* function of v that will numerically solve the differential equation (3.ii) using initial conditions $\theta(0) = 0, \theta'(0) = v$. Do likewise for equation (3.iii). Plot the solutions to both the nonlinear equation (3.ii) and the linear equation (3.iii) for each of the following values of the initial velocity v: 1, 1.99, 2, 2.01. Use a time interval from $t = 0$ to $t = 40$. Clearly mark which is the nonlinear solution and which is the linear solution on each plot. Compare the linear and nonlinear behavior in each case. For the nonlinear equation, interpret what the graph indicates the pendulum is doing physically. Speculate as to the true motion.

5. In this problem, we'll investigate the effect of damping on the pendulum, using the model

$$\theta'' + b\theta' + \sin\theta = 0$$

(*cf.* Problem 3). Define a *Maple* function of b that will find a numerical solution of this differential equation with initial conditions $\theta(0) = 0, \theta'(0) = 4$. Then plot the solution from $t = 0$ to $t = 20$ for the values $b = 0.5, 1,$ and 2. Interpret what is happening *physically* in each case, *i.e.*, describe explicitly what the graph says the pendulum is doing. Now do the same for the linear approximation

$$\theta'' + b\theta' + \theta = 0,$$

for the same values of b. Compare the linear and nonlinear behavior in each case.

6. In this problem, we'll look at the effect of a periodic external force on the pendulum, using the model

$$\theta'' + 0.05\theta' + \sin\theta = 0.3\cos\omega t$$

(*cf.* Problems 3–5). We have chosen a value for the damping coefficient that is more typical of air resistance than the values in the previous problem. Write a

Maple function of w that numerically solves this equation with initial conditions $\theta(0) = 0, \theta'(0) = 0$. Then plot the result from $t = 0$ to $t = 60$ for the following values of the frequency w: $0.6, 0.8, 1, 1.2$. Which frequency moves the pendulum farthest away from its equilibrium position? Do the same analysis for the linear approximation

$$\theta'' + 0.05\theta' + \theta = 0.3\cos wt$$

and answer the same question. For which frequencies do the linear and nonlinear equations have widely different behaviors? Which forcing frequency seems to induce resonance-type behavior in the pendulum? Graph that solution on a longer interval and decide whether the amplitude goes to infinity.

7. In many applications, second order equations come with *boundary conditions* rather than *initial conditions*. For example, consider a cable that is attached at each end to a post, with both ends at the same height (let us call this height $y = 0$). If the posts are located at $x = 0$ and $x = 1$, the height y of the cable as a function of x satisfies the differential equation

$$y'' = c\sqrt{1 + (y')^2}$$

with boundary conditions $y(0) = 0, y(1) = 0$. The constant c depends on the length of the cable; for this problem we'll use $c = 1$. *Maple* can't quite solve this boundary problem. Nonetheless, we can approximate the solution of the differential equation that satisfies the boundary value conditions as follows. For a given value of s, find a solution to the above equation (using $c = 1$) with initial conditions $y(0) = 0, y'(0) = s$. Find the value of s that leads to a solution satisfying the condition $y(1) = 0$. Graph the solution on $[0, 1]$ and determine the maximum dip in the cable.

8. The problem of finding the function $u(x)$ satisfying

$$\begin{cases} a(x)u''(x) + a'(x)u'(x) = f(x), 0 \le x \le 1 \\ u(0) = u(1) = 0 \end{cases}$$

arises in studying the longitudinal displacements in a longitudinally loaded elastic bar. The bar is of length 1, its left end is at $x = 0$, and its right end at $x = 1$. In the differential equation above, $f(x)$ represents the external force on the bar (which is assumed to be longitudinal, *i.e.*, directed along the bar), $a(x)$ represents both the elastic properties and the cross-sectional area of the bar, and $u(x)$ is the longitudinal displacement of the bar at the point x. This problem is an example of a *boundary value problem*; the conditions $u(0) = u(1) = 0$, which specify that the ends of the bar are fixed, are called boundary conditions. The function $a(x)$ may be constant; this is the case if neither the elastic properties nor the cross-sectional area depends on the position x in the rod, *i.e.*, if the rod is uniform. But if the rod is not uniform, then $a(x)$ is not constant, and the equation has variable coefficients. Similarly, $f(x)$ will be constant only if the external force is applied uniformly along the rod.

Now suppose $a(x) = \pi(2 + \cos^2(2\pi x))$ and $f(x) = 1500x^4$. Find and plot the corresponding solution $u(x)$. Since the equation has variable coefficients, constant coefficient methods are not applicable, so we will use a numerical method. We cannot apply **dsolve(..., numeric)** directly, since it requires initial conditions. Nonetheless, we can use **dsolve(..., numeric)** as follows to approximate the solution. Write a function that, for a given value of s, finds a numerical solution to the equation with the initial conditions $u(0) = 0, u'(0) = s$ and plots this solution for $0 \le x \le 1$. Note that the first boundary condition is one of our initial conditions. Then by evaluating this function for various values of s you should be able to get a fairly accurate value for the s that leads to a solution satisfying the second boundary condition $u(1) = 0$. Next define a second function (of s) that finds a numerical solution of the equation with the initial conditions $u(0) = 0, u'(0) = s$, and then numerically evaluates $u(1)$. Using this function, you should be able to get an accurate approximation for the desired s. Plot the final (accurate) solution to the boundary value problem. What is the maximum displacement, and where does it occur?

9. In this problem, we study the effects of air resistance.

(a) A paratrooper steps out of an airplane at a height of 1000 ft, and after 5 seconds opens her parachute. Her weight, with equipment, is 195 lbs. Let $y(t)$ denote her height above the ground after t seconds. Assume that the air resistance is $0.005y'(t)^2$ lbs in free fall and $0.6y'(t)^2$ lbs with the chute open. At what height does the chute open? How long does it take to reach the ground? At what velocity does she hit the ground? (This model assumes that air resistance is proportional to the *square* of the velocity, and that the parachute opens instantaneously.)

(*Hint*: Pay attention to units. Recall that the mass of the paratrooper is $195/32$, measured in lb sec^2/ft. Here, 32 is the acceleration due to gravity, measured in ft/sec^2.)

(b) Let $v = y'$ be the velocity during the second phase of the fall (while the chute is open). One can view the equation of motion as an autonomous first order ODE in the velocity:

$$v' = -32 + \frac{192}{1950}v^2.$$

Make a qualitative analysis of this first order equation, finding in particular the critical or equilibrium velocity. This velocity is called the terminal velocity. How does the terminal velocity compare with the velocity at the time the chute opens and with the velocity at impact?

(c) Assume the paratrooper is safe if she strikes the ground at a velocity within 5% of the terminal velocity in (b). Except for the initial height, use the parameters in (a). What is the lowest height from which she may parachute safely? (*Please do not try this at home!*)

10. This problem is based on Problems 25 and 26 in Boyce & DiPrima, Section 3.8.
Consider the IVP

$$u'' + cu' + u = 0, \quad u(0) = 2, \quad u'(0) = 0.$$

(a) Find the solution. For $c = 0.25$, graph the solution. Determine an approximate time τ at which the solution becomes "negligible." Interpret negligible to mean that no meaningful difference between the curve and the x-axis can be discerned on your plot. Repeat this process for the values $c = 0.25, 0.5, \ldots, 1.75, 2$. Make a list of values (c, τ) corresponding to the eight values of c. Plot the resulting points. How does τ depend on c? What does the theory of damped vibrations predict about the relationship between τ and c?

(b) Now plot the solution curves for the values $c = 1.75, 1.8, \ldots, 2.2, 2.25$. The issue we would like to resolve is: at what point does the transition from oscillatory to nonoscillatory behavior occur. The characteristic equation tells you it is $c = 2$. Why? By adjusting your domain and range, try to verify that. If you cannot, explain what prevents you from doing so. (*Hint:* For $c = 1.9$, for example, compute where the curve first crosses the x-axis. What is the magnitude at nearby points?)

11. This problem is based on Problems 18 and 19 in Boyce & DiPrima, Section 3.9.
Consider the IVP

$$u'' + u = 3\cos(\omega t), \quad u(0) = 0, \quad u'(0) = 0.$$

(a) Find the solution (using **dsolve**). For $\omega = 0.5, 0.6, 0.7, 0.8, 0.9$, plot the solution curves on the interval $0 \le x \le 10$. Note that $\omega_0 = 1$ is the natural frequency of the homogeneous equation. Describe how the solution curves behave as ω gets closer to 1.

(b) Notice that the formula in (a) is invalid when $\omega = 1$. Find and plot the solution curve for $\omega = 1$. What phenomenon is exhibited? Corroborate your conclusion by plotting on a longer interval.

(c) Compare the plots for $\omega = 1$ and $\omega = 0.9$ on a longer interval to see that the behavior for $\omega = 0.9$ is different from that in (b). What phenomenon is exhibited by the curve for $\omega = 0.9$?

12. In this problem we study how solutions of the initial value problem

$$\frac{d^2y}{dt^2} + 0.15\frac{dy}{dt} - y + y^3 = 0, \quad y(0) = a, \quad y'(0) = 0$$

depend on the initial value a.

Plot a numerical solution of this equation from $t = 0$ to $t = 40$ for each of the initial values $a = 0.5, 1, 1.5, 2, 2.5$. You may have to adjust **maxfun** and **stepsize** in order to make some of these plots.

Describe how increasing the initial value of y affects the solutions, both in terms of their limiting behavior and their general appearance. (In making comparisons, pay careful attention to the scale on the y-axis.)

Finally, plot all five solutions on one graph. Would such a picture be possible for solutions of a first order differential equation? Why or why not?

13. In this problem, we consider the long-term behavior of solutions of the initial value problem

$$\frac{d^2y}{dt^2} + 0.15\frac{dy}{dt} - y + y^3 = 0.3\cos(\omega t), \quad y(0) = 0, \quad y'(0) = 0$$

for various frequencies ω in the forcing term.

Plot a numerical solution of this equation from $t = 0$ to $t = 100$ for each of the eight frequencies $\omega = 0.8, 0.9, \ldots, 1.4, 1.5$. You may have to adjust **maxfun** and **stepsize** in order to make these plots. You should also expect *Maple* to take more time than usual in making them.

Describe and compare the different long-term behaviors you see. Due to the forcing term, all solutions will oscillate, but pay particular attention to the magnitude of the oscillations, and to whether or not there is a periodic pattern to them. Are there any similarities between your results for this nonlinear system and the phenomenon of resonance for linear systems with periodic forcing?

14. In this problem, we study the zeros of solutions of the second order differential equation

$$y'' + (2 + \sin x)y = 0. \tag{i}$$

(a) Compute and plot several solutions of this equation with different initial conditions: $y(0) = c, y'(0) = d$. To be specific, choose three different values for the pair c, d and plot the corresponding solutions over $[0, 20]$. By inspecting your plots, find a number L that is an upper bound for the distance between successive zeros of the solutions. Then find a number l that is a lower bound for the distance between successive zeros.

(b) Information on the zeros of solutions of linear second order ODEs can be obtained from the Sturm Comparison Theorem (see Chapter 9). By comparing (i) with the equation $y'' + y = 0$, you should be able to get a value for L, and by comparing (i) with $y'' + 3y = 0$, you should be able to find l. How do these values compare with the values obtained in part (a)? (You will find it useful to note that the general solution of $y'' + ky = 0$, where k is a positive constant, can be written as $y = R\cos(\sqrt{k}x - \delta)$, with $R \geq 0, \delta$ arbitrary. R is called the *amplitude* and δ the *phase shift*.)

(c) Plot a solution of (i) and a solution of $y'' + y = 0$ on the same graph, and verify that between any two zeros of the latter solution, there is at least one zero of the solution of (i).

15. Consider Bessel's equation of order zero

$$x^2 y'' + x y' + x^2 y = 0.$$

(a) Compute and plot several solutions of this equation with different initial conditions: $y(1) = c, y'(1) = d$. To be specific, choose three different values for the pair c, d and plot the corresponding solutions over $[0.1, 20]$. (Why isn't it a good idea to use $x = 0$ in the initial conditions?) By inspecting your plots, find a number L that is an upper bound for the distance between successive zeros of the solutions. Then find a number l that is a lower bound for the distance between successive zeros.

(b) Now confirm your findings with the Sturm Comparison Theorem. It doesn't apply directly, but if we introduce the new function $z(x) = x^{1/2} y(x)$, Bessel's equation becomes

$$z'' + (1 + 1/(4x^2))z = 0,$$

which has the form of the equation in the Comparison Theorem. Since y will have a zero wherever z has a zero, we can study the zeros of y by studying the zeros of z. By comparing the equation for z with the equation $z'' + z = 0$, determine an upper bound L on the distance between successive zeros of any solution of Bessel's equation.

(c) Let a be a positive number. For $x \in [a, \infty)$, the quantity $1 + 1/(4x^2)$ is less than or equal to the constant $1 + 1/(4a^2)$. By making an appropriate comparison, determine a lower bound l on the distance between successive zeros of any solution of Bessel's equation (for $x > a$). What is the limiting value of l as a goes to ∞? Approximately how far apart are the zeros when x is large? Did your graphical study lead to comparable values for l and L?

16. A solution of a second order linear differential equation is called *oscillatory* if its graph crosses the x-axis infinitely many times, and *nonoscillatory* if it crosses only finitely many times. In this problem, we will be interested in determining the oscillatory nature of nonzero solutions to some second order linear ODEs. First consider the equation $y'' + ky = 0$, where k is a constant. If $k > 0$, it has the general solution $y = R\cos(\sqrt{k}x - \delta)$, with $R \geq 0, \delta$ arbitrary. R is called the *amplitude* and δ the *phase shift*. From this formula we see that any solution $y(x)$ has infinitely many zeros and hence crosses the axis infinitely many times, *i.e.*, is oscillatory. If $k < 0$, the general solution is $y(x) = c_1 e^{\sqrt{-k}x} + c_2 e^{-\sqrt{-k}x}$, so we see that solutions have at most one zero, *i.e.*, are nonoscillatory. If $k = 0$, the general solution is $y(x) = c_1 x + c_2$. These have at most one zero and so are nonoscillatory.

Next consider Airy's equation

$$y'' = xy, \tag{i}$$

which arises in various applications. Since (i) has a variable coefficient, we cannot study it by elementary methods; in particular, we cannot find the general solution as we did for the constant coefficient equation. We will instead first make a graphical study of the solutions of (i), and then study the solutions using the Sturm Comparison Theorem (see Chapter 9). Compute and plot several numerical solutions of (i) with different initial conditions $y(0) = c, y'(0) = d$. To be specific, choose three different values for the pair c, d, and graph the corresponding solutions over the interval $[-10, 5]$. What do you observe? Do solutions have zeros on the negative x-axis? A few? Many? Do they get closer together as x becomes more negative? Do solutions have zeros on the positive x-axis? Where are the solutions oscillatory?

Information on the zeros of solutions of linear second order ODEs, and hence information on their oscillatory nature, can also be obtained from the Sturm Comparison Theorem. By comparing (i) with the equation $y'' - by = 0$ for $x \leq b < 0$, what do you learn about zeros on the negative x-axis and their spacing, and hence about the oscillatory nature of solutions of (i) for $x \leq 0$? By comparing (i) with $y'' = 0$ for $0 < x$, what do you learn about the oscillatory nature of solutions for $x > 0$? How many zeros could a solution have on the positive x-axis? Do the graphs you plotted above confirm these results?

17. In this problem, we study solutions of the parabolic cylinder equation

$$y'' + \left(n + \frac{1}{2} - \frac{x^2}{4}\right) y = 0,$$

which arises in the study of quantum-mechanical vibrations. Since the equation is unchanged if x is replaced by $-x$, any solution function will be symmetric with respect to the y-axis. Therefore, we focus our attention on $x \geq 0$.

(a) Find the corresponding first order equation for $z = \arctan(y/y')$, as described in the *Geometric Method* section of Chapter 9.

(b) For $n = 1$, plot the direction field for the z equation from $x = 0$ to $x = 10$. (Remember to use $-\pi/2 \leq z \leq \pi/2$.) Based on the plot, predict what the solutions y to the parabolic cylinder equation look like near $x = 0$, and for larger x. Is there a value of x around which you expect their behavior to change?

(c) Now numerically solve the parabolic cylinder equation for $n = 1$ with two sets of initial conditions $y(0) = 1, y'(0) = 0$ and $y(0) = 0, y'(0) = 1$. Plot the two solutions on the same graph. (It will probably help to change the range on the plot.) Do the solutions behave as you expected?

(d) Repeat parts (b) and (c) for $n = 5$. Point out any differences from the case $n = 1$.

(e) Repeat parts (b) and (c) for $n = 15$. Discuss how the solutions are changing as n increases.

(f) Now consider the three solutions (for $n = 1, 5,$ and 15) with $y(0) = 0$. Point out similarities and differences. By drawing an analogy with Airy's equation, argue from the direction fields that, for any n, exactly one solution function decays for large x while all others grow. Do you have enough graphical evidence from the numerical plots to conclude that the solution corresponding to the initial data $y(0) = 0$, $y'(0) = 1$ is that function? Why or why not? (*Hint:* Look to previous discussions of stability of solutions as a guide.)

18. In this problem, we study solutions of Bessel's equation of order n,

$$y'' + \frac{1}{x}y' + \left(1 - \frac{n^2}{x^2}\right)y = 0,$$

for $n > 0$. Solutions of this equation, called Bessel functions of order n, are used in the study of vibrations and waves with circular symmetry. Since the equation is unchanged if x is replaced by $-x$, we focus our attention on $x \geq 0$.

(a) Find the corresponding first order equation for $z = \arctan(y/y')$, as described in the *Geometric Method* section of Chapter 9.

(b) For $n = 1$, plot the direction field for the z equation from $x = 0$ to $x = 20$. (Remember to use $-\pi/2 \leq z \leq \pi/2$.) Based on the plot, predict what the Bessel functions of order 1 look like near $x = 0$, and for larger x. Is there a value of x around which you expect their behavior to change?

(c) Now plot the Bessel functions **BesselJ(n, x)** and **BesselY(n, x)** for $n = 1$ on the same graph. (It will probably help to change the range on the plot.) Do the solutions behave as you expected?

(d) Repeat parts (b) and (c) for $n = 5$. Point out any differences from the case $n = 1$.

(e) Repeat parts (b) and (c) for $n = 15$. Discuss how the solutions are changing as n increases.

Chapter 10

Series Solutions

A primary theme of this book is that numerical, geometric, and qualitative methods can be used to study solutions of differential equations, even when we cannot find an exact formula solution. Of course, exact solutions are extremely valuable, and there are many techniques for finding them. For instance, techniques for finding exact solutions of second order linear differential equations include:

- The exponential substitution, which leads to solutions of an arbitrary constant coefficient homogeneous equation;
- The method of *reduction of order*, which produces a second linearly independent solution of a homogeneous equation when one solution is already known;
- The method of *undetermined coefficients*, which solves special kinds of inhomogeneous equations with constant coefficients; and
- The method of *variation of parameters*, which yields the solution of a general inhomogeneous equation, given a fundamental set of solutions to the homogeneous equation.

Each of these techniques reduces the search for a formula solution to a simpler problem in algebra or calculus: finding the root of a polynomial, computing an antiderivative, or solving a pair of simultaneous linear equations.

The process of finding formulas for exact solutions of equations, either by hand or by computer, is called *symbolic computation*. The **dsolve** command incorporates some of the techniques listed above. It enables us to find exact solutions more rapidly and more reliably than we could by hand.

There are many differential equations that do not yield to the techniques listed above. By using more advanced ideas from calculus, however, we can find exact solutions for a wider class of differential equations. In this chapter and the next, we discuss two such calculus-based techniques for finding exact solutions of differential equations: *Series Solutions* and *Laplace Transforms*. The method of Series Solutions enables us to solve differential equations with variable coefficients. The theory of Laplace Transforms (discussed in the next chapter) enables us to solve constant coefficient linear differential equations with discontinuous inhomogeneous terms. These are sophisticated techniques, involving improper

135

integrals, complex variables, and power series. Both techniques are computation-
ally intensive, and both have been implemented as options to **dsolve**.

Series Solutions

Consider the second order homogeneous linear differential equation with variable
coefficients

$$a(x)y''(x) + b(x)y'(x) + c(x)y(x) = 0. \tag{1}$$

Suppose the coefficient functions a, b, and c are *analytic* (*i.e.*, have convergent
power series representations) in a neighborhood of a point $x = x_0$. For simplicity,
we shall assume that $x_0 = 0$, but everything we say remains valid for any point of
analyticity. We begin by dividing by $a(x)$ to normalize the equation. This causes
no problems as long as $a(0) \neq 0$. In this case, we refer to the origin as an *ordinary
point* for the equation. Changing notation, we write

$$y''(x) + p(x)y'(x) + q(x)y(x) = 0. \tag{2}$$

Equations (1) and (2) have the same solutions. We will search for a power series
solution of the form

$$y(x) = \sum_{n=0}^{\infty} a_n x^n. \tag{3}$$

If we can find such a solution, then $y(0) = a_0$ and $y'(0) = a_1$. Thus, if we are
given initial values, then we can find the first two coefficients of the power series
solution.

By substituting the infinite series (3) into equation (2), expanding p and q as
power series, and combining terms of the same degree, we obtain a *recursion
relation* for the coefficients a_n. In other words, we obtain algebraic equations for
a_n that involve $a_0, a_1, \ldots, a_{n-1}$ and known coefficients of the power series of p
and q. Sometimes, we can solve the recursion relation in *closed form* by finding an
algebraic formula for a_n. This formula would involve known functions of n (such
as $n!$ or powers of n) and a_0 and a_1, but no other coefficients of lower degree. If no
initial data are given, then a solution to the recursion relation produces a general
solution of the differential equation with arbitrary constants a_0 and a_1.

Solving a recursion relation in closed form is like finding an exact formula
solution for a differential equation. In many cases, we cannot solve the recursion
relation. Nevertheless, we can still use the recursion relation to compute as many
coefficients as we wish. As long as we stay close to the origin, the leading terms
should give a good approximation to the full power series solution. One must be
careful, however, because the power series solution will be valid only inside its
radius of convergence. The radius of convergence will be at least as large as the
distance from the origin to the nearest singularity of p or q. When the recursion
relation is solvable in closed form, we can use a standard test from calculus to
compute the radius of convergence precisely. In general, we can only make an
educated guess.

We can use *Maple* to automate the process of finding power series solutions. In fact, the power series method is built into the **dsolve** command and is invoked with the option **type = series**.

EXAMPLE. Consider the initial value problem

$$y'' - xy' - y = 0, \qquad y(0) = 2, \qquad y'(0) = 1$$

from Problem 15 in Section 5.2 of Boyce & DiPrima. We can solve this with **dsolve** by typing

eqn1 := diff(y(x), x$2) − x*diff(y(x), x) − y(x) = 0;
sol1 := dsolve({eqn1, y(0) = 2, D(y)(0) = 1}, y(x), type = series);

The resulting power series is

$$y(x) = 2 + x + x^2 + \frac{1}{3}x^3 + \frac{1}{4}x^4 + \frac{1}{15}x^5 + O\left(x^6\right).$$

The symbol $O(x^6)$ refers to all terms of order at least 6. If we drop the term $O(x^6)$ from this expression, we get a *Taylor polynomial*, which approximates the power series. In *Maple*, we do this by typing **convert(rhs(sol1), polynom)**.

The order of the series solution is determined by the global variable **Order**, whose default value is 6. To see the series solution to a different order we redefine **Order**. For example, to see all the terms of the series solution up to order 10, we type

Order := 11;
dsolve({eqn1, y(0) = 2, D(y)(0) = 1}, y(x), type = series);

which yields

$$y(x) = 2 + x + x^2 + \frac{1}{3}x^3 + \frac{1}{4}x^4 + \frac{1}{15}x^5 + \frac{1}{24}x^6 + \frac{1}{105}x^7$$
$$+ \frac{1}{192}x^8 + \frac{1}{945}x^9 + \frac{1}{1920}x^{10} + O\left(x^{11}\right).$$

The initial conditions tell *Maple* where to expand the power series. In the preceding examples, the initial conditions were given at $x = 0$, so the series solution was given as an expansion at $x = 0$. To expand at $x = 1$, for example, we could type

dsolve({eqn1, y(1) = 1, D(y)(1) = 0}, y(x), series);

The result will be a power series expansion at $x = 1$ of the solution of the IVP; *i.e.*, a power series in $(x - 1)$.

The **dsolve** command without the **series** option will find exact solutions to many homogenous second order equations with polynomial coefficients. Often, these solutions involve special functions like **BesselJ** and **BesselY**. Equations with coefficients that are not polynomials generally require series methods.

Singular Points

In the discussion above, we considered equations of the form

$$a(x)y'' + b(x)y' + c(x)y = 0,$$

where $a(0) \neq 0$. We transformed the equation into the form

$$y'' + p(x)y' + q(x)y = 0$$

by dividing by $a(x)$. If $a(0) = 0$, then dividing by $a(x)$ may result in one or both of $p(x)$ and $q(x)$ being singular at $x = 0$. In this case, we say that the equation has a *singularity* at $x = 0$.

A prototype is the Euler equation:

$$x^2 y'' + bxy' + cy = 0,$$

where b and c are constants. Dividing by x^2 yields the equation

$$y'' + \frac{b}{x}y' + \frac{c}{x^2}y = 0,$$

so $p(x) = b/x$ and $q(x) = c/x^2$. Both $p(x)$ and $q(x)$ are singular at $x = 0$. Such isolated singularities, where the singularity of p is no worse than $1/x$ and the singularity of q is no worse than $1/x^2$, are called *regular singular points*. More precisely, we say that a function $f(x)$ has a *pole of order* n at x_0 if $\lim_{x \to x_0} f(x) = \infty$, and n is the smallest integer such that $\lim_{x \to x_0}(x - x_0)^n f(x)$ is finite. A homogeneous linear differential equation $y^{(n)} + p_1(x)y^{(n-1)} + \ldots + p_n(x)y = 0$ is said to have a regular singular point at x_0 if it is singular at x_0, and p_k has a pole of order at most k at x_0.

Here is a summary of the general approach to solving second order linear equations with regular singular points. We suppose, for simplicity, that the singular point is $x = 0$, and we look for solutions on the interval $(0, \infty)$. Let $p_0 = \lim_{x \to 0} xp(x)$ and $q_0 = \lim_{x \to 0} x^2 q(x)$. Let r_1 and r_2 be the roots of the *indicial equation* $r(r - 1) + p_0 r + q_0 = 0$. We look for a solution of the differential equation of the form

$$y_1(x) = x^{r_1} u(x),$$

where $u(0) = 1$. If $r_1 - r_2$ is not an integer, then we look for a second solution of the form

$$y_2(x) = x^{r_2} v(x),$$

where $v(0) = 1$. If $r_1 = r_2$, then we look for a second solution of the form

$$y_2(x) = y_1(x)\ln(x) + x^{r_1} v(x),$$

where $v(0) = 0$. If $r_1 - r_2$ is a positive integer, then we look for a solution of the form

$$y_2(x) = ay_1(x)\ln(x) + x^{r_1} v(x),$$

where a is a constant and $v(0) = 1$. The functions $u(x)$ and $v(x)$ will be analytic at 0, and therefore will have power series expansions at $x = 0$. These solutions are called *Frobenius series* solutions.

One could use *Maple* to apply these techniques by hand, but **dsolve(. . ., series)** already uses them to solve equations with regular singular points. When applying **dsolve** to singular equations, it is important to remember that the solutions may blow up at the singular point. Consequently, it often does not make sense to specify an initial condition at the singular point.

EXAMPLE. Consider the differential equation:

$$2xy'' + y' + xy = 0.$$

The origin is a regular singular point, and the indicial equation has roots $r = 0$ and $r = 1/2$. Hence, the differential equation should have a Frobenius series solution of the form

$$x^{1/2} \sum_{n=0}^{\infty} a_n x^n.$$

To solve this equation with **dsolve**, we type:

eqn2 := 2*x*diff(y(x), x\$2) + diff(y(x), x) + x*y(x) = 0;
dsolve(eqn2, y(x), type = series);

The solution is:

$$y(x) = \_C1 \sqrt{x} \left(1 - \frac{1}{10}x^2 + \frac{1}{360}x^4 + O\left(x^6\right)\right)$$
$$+ \_C2 \left(1 - \frac{1}{6}x^2 + \frac{1}{168}x^4 + O\left(x^6\right)\right).$$

Note that we did not specify an initial condition in this example, although we could have since the solutions are nonsingular at 0. In general, specifying an initial condition (at the singular point) for a singular differential equation will cause **dsolve** to fail.

EXERCISE. Use **dsolve** with the **series** option to solve the Euler equation

$$x^2y'' + 3xy' + y = 0.$$

Are the solutions singular at 0? What happens if you specify an initial condition at $x = 0$ in the **dsolve** command?

It is important to note that for equations with regular singular points, the Frobenius series tells us how fast the solution blows up at the singularity, *e.g.*, the solution blows up like $1/x$, or $x^{-1/2}$, *etc.* This information is not easily gleaned from a numerical solution. For equations with an irregular singular point, we cannot expect a Frobenius series solution to be valid, and other techniques (numerical or qualitative) must be used. Some of those techniques are addressed in Problem Set E.

In this chapter, we have focused on linear homogeneous second order differential equations. The method of series solutions can also be used for inhomogeneous equations, higher order equations, and nonlinear equations. In fact, as we've noted before, most differential equations cannot be solved in terms of elementary functions. For many of these equations, a series solution is the only analytic solution available.

Chapter 11

Laplace Transforms

A *transform* is a mathematical operation that changes a given function into a new function. Transforms are often used in mathematics to change a difficult problem into a more tractable one. In this chapter, we introduce the *Laplace Transform*, which is particularly useful for solving linear differential equations with constant coefficients and discontinuous inhomogeneous term. The key feature of the Laplace Transform is that (roughly speaking) it changes the operation of differentiation into the operation of multiplication. Thus the Laplace Transform changes a differential equation into an algebraic equation. To solve a linear differential equation with constant coefficients, you apply the Laplace Transform to change the differential equation into an algebraic equation, solve the algebraic equation, and then apply the Inverse Laplace Transform to transform the solution of the algebraic equation back into a solution of the differential equation.

The Laplace Transform of a function f is a new function, denoted by F or by $\mathcal{L}(f)$, and defined as follows:

$$F(s) = \mathcal{L}(f)(s) = \int_0^\infty f(t)e^{-st}\, dt.$$

This transform is called an *integral transform* because it is obtained by integrating the function f against another function e^{-st}, called the *kernel* of the transform. The integral in question is an improper integral, so we have to make sure that it converges. Note that while the argument s of the function F appears in the integrand, the integration is with respect to the variable t. Note also that the integral is over the domain $[0, \infty)$, so we need only assume that the function f is defined for $t \geq 0$.

To get a feel for the Laplace Transform, we compute the Laplace Transform of the function e^{at}.

$$\mathcal{L}(e^{at})(s) = \int_0^\infty e^{at}e^{-st}\, dt = \lim_{c \to \infty} \int_0^c e^{at}e^{-st}\, dt$$

$$= \lim_{c \to \infty} \int_0^c e^{(a-s)t}\, dt = \lim_{c \to \infty} \left. \frac{e^{(a-s)t}}{(a-s)} \right|_0^c =$$

$$= \lim_{c \to \infty} \left(\frac{e^{(a-s)c}}{(a-s)} - \frac{1}{a-s} \right)$$

$$= \begin{cases} 1/(s-a), & \text{if } s > a \\ +\infty, & \text{if } s \leq a. \end{cases}$$

Note that the Laplace Transform of e^{at} is defined for $s > a$. In fact, a straightforward argument shows that if f is any piecewise continuous function on $[0, \infty)$ with the property that $|f(t)| \leq Ke^{at}$, for some constant $K > 0$, then the improper integral defining the Laplace Transform converges for $s > a$, and therefore $\mathcal{L}(f)(s)$ is defined for $s > a$. We say that a function is *piecewise continuous* if it only has a discrete set of jump discontinuities. If f satisfies an inequality of the form $|f(t)| \leq Ke^{at}$, we say that f is of *exponential order*. Most functions one encounters in practice are of exponential order. From now on, we only consider piecewise continuous functions of exponential order. In particular, if f is a bounded function, then it satisfies the inequality $|f(t)| \leq K = Ke^{0t}$ for some $K > 0$, so $\mathcal{L}(f)(s)$ is defined for all $s > 0$. More generally, any function whose growth is of exponential order has a Laplace Transform which is defined for sufficiently large s.

EXERCISE. Compute the Laplace Transform of the functions $f(t) = 1$ and $\cos t$. (For $\cos t$, you must integrate by parts twice.)

We asserted that the Laplace Transform transforms differentiation into multiplication. This is a consequence of the integration by parts formula:

$$\mathcal{L}(f')(s) = \int_0^\infty f'(t)e^{-st}\, dt = \lim_{c \to \infty} \int_0^c f'(t)e^{-st}\, dt$$

$$= \lim_{c \to \infty} \left[f(t)e^{-st}\big|_0^c - \int_0^c f(t)(-se^{-st})\, dt \right] \qquad (1)$$

$$= -f(0) + s \lim_{c \to \infty} \int_0^c f(t)e^{-st}\, dt$$

$$= s\mathcal{L}(f)(s) - f(0).$$

To be precise, we must assume that f is differentiable and f' is piecewise continuous for this formula to hold. We can summarize (1) as follows: If the Laplace Transform of f is $F(s)$, then the Laplace Transform of f' is $sF(s) - f(0)$. In other words, the Laplace Transform changes the operation of differentiation into the operation of multiplication (by the independent variable) plus a translation (by $-f(0)$). The formula in (1) has a straightforward generalization to higher derivatives. If the Laplace Transform of f is F, $f^{(k)}$ is continuous for $k = 0 \ldots n-1$, and $f^{(n)}$ is piecewise continuous, then

$$\mathcal{L}(f^{(n)})(s) = s^n F(s) - s^{n-1}f(0) - s^{n-2}f'(0) - \cdots - f^{(n-1)}(0).$$

The Laplace Transform has many other important properties, of which we mention two here. First, the Laplace Transform is linear, *i.e.*, $\mathcal{L}(af+bg) = a\mathcal{L}(f)+b\mathcal{L}(g)$.

Linearity of the Laplace Transform follows easily from linearity of integration. Second, the Laplace Transform is invertible; the inverse is called the Inverse Laplace Transform, and is denoted by $\mathcal{L}^{-1}$. The Inverse Laplace Transform has the property that $\mathcal{L}^{-1}(\mathcal{L}(f)) = f$, *i.e.*, the Inverse Laplace Transform of the Laplace Transform of a function is the function itself. The Inverse Laplace Transform is also an integral transform, but it involves contour integrals in the complex plane, so we do not give the definition here.

Solving Differential Equations with Laplace Transforms

Let's see what happens when we apply the Laplace Transform to a second order linear differential equation with constant coefficients. Consider the initial value problem

$$ay''(t) + by'(t) + cy(t) = f(t), \qquad y(0) = y_0, \qquad y'(0) = y_0'. \tag{2}$$

When we apply the Laplace Transform to this equation, we get the algebraic equation

$$a(s^2 Y(s) - sy_0 - y_0') + b(sY(s) - y_0) + cY(s) = F(s),$$

where $Y(s)$ is the Laplace Transform of $y(t)$, and $F(s)$ is the Laplace Transform of $f(t)$. We solve this algebraic equation for $Y(s)$, to get

$$Y(s) = \frac{F(s) + asy_0 + ay_0' + by_0}{as^2 + bs + c},$$

and then compute the Inverse Laplace Transform of the right-hand side to get an expression for $y(t)$.

Traditionally, Laplace Transforms and Inverse Laplace Transforms were computed by reading tables. You can also use *Maple* to compute Laplace Transforms. First, you must load the Laplace Transform commands from the integral transforms package by typing **with(inttrans)**. The Laplace Transform of a function $y(t)$ is then computed by

    laplace(y(t), t, s);

and the inverse transform of a function $Y(s)$ is computed by

    invlaplace(Y(s), s, t);

For example, **laplace(cos(t), t, s)** yields $s/(1+s^2)$.

You could use the **laplace** and **invlaplace** commands to solve differential equations in *Maple*; but *Maple* has already implemented the Laplace Transform method as an option to the **dsolve** command. Thus we could solve the initial value problem (2) by typing

    eqn1 := a*diff(y(t), t$2) + b*diff(y(t), t) + c*y(t) = f(t);
    dsolve({eqn1, y(0) = y0, D(y)(0) = yp0}, y(t), method = laplace);

Of course, without particular values for a, b, c, f, y_0 and y_0', the solution is a complicated symbolic expression.

EXAMPLE. Consider the initial value problem:

$$y''(t) - 2y(t) = \sin 2t, \quad y(0) = 1, \quad y'(0) = 2.$$

Typing

> **eqn2 := diff(y(t), t\$2) − 2*y(t) = sin(2*t);**
> **dsolve({eqn2, y(0) = 1, D(y)(0) = 2}, y(t), method = laplace);**

produces the solution

$$y(t) = \frac{7}{6}\sqrt{2}\sinh(\sqrt{2}t) + \cosh(\sqrt{2}t) - \frac{1}{6}\sin(2t).$$

The **method = laplace** option is not necessary to solve this initial value problem (although it makes the solution look nicer). In fact, without this option, **dsolve** uses the method of variation of parameters. If you were solving the IVP by hand, you could also use the method of undetermined coefficients. The full power of the Laplace Transform method comes into play when solving equations that involve discontinuous functions.

Discontinuous Functions

The Laplace Transform is especially useful for solving differential equations that involve piecewise continuous functions. The basic building block for piecewise continuous functions is the *unit step function u(t)*, defined by

$$u(t) = \begin{cases} 0, & \text{if } t < 0, \\ 1, & \text{if } t \geq 0. \end{cases}$$

In honor of the physicist and engineer Oliver Heaviside (1850-1925), who developed the Laplace Transform method to solve problems in electrical engineering, this function is sometimes called the Heaviside function. In *Maple*, this function has been implemented with the name **Heaviside**.

The unit step function is best thought of as a switch, which is off until time 0, and then on starting at time 0. The function $1 - u(t)$ is also a switch, on until time 0, and then off. To make a switch that comes on at time c, we simply translate $u(t)$ by c; thus $u(t - c)$ is a switch that comes on at time c. Similarly, $1 - u(t - c)$ is a switch that goes off at time c. The unit step function translated by c is often written $u_c(t) = u(t - c)$. Note that $1 - u(t-c) = u(c-t)$, or equivalently, $1 - u_c(t) = u_{-c}(-t)$.

The unit step function can be used to build piecewise continuous functions by switching pieces of the function on and off at appropriate times. To turn a function on at time c, multiply it by $u_c(t)$, and to turn it off at time d, multiply it by $(1-u_d(t))$. Consider, for example, the function

$$f(t) = \begin{cases} 0, & t < 0, \\ 1, & 0 \leq t < 1, \\ t^2, & 1 \leq t < 3, \\ \sin 2t, & t \geq 3. \end{cases}$$

We can write this as

$$f(t) = u_0(t)(1 - u_1(t)) + u_1(t)(1 - u_3(t))\, t^2 + u_3(t)\sin 2t.$$

In *Maple*, we would enter this as

    f := t –> Heaviside(t)*Heaviside(1 – t)
        + Heaviside(t – 1)*Heaviside(3 – t)*t^2
        + Heaviside(t – 3)*sin(2*t)

Here, we have used the identity **Heaviside(c – t) = (1 – Heaviside(t – c))**. Here is a plot of f.

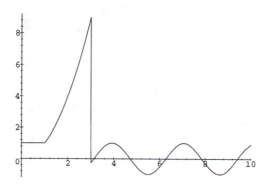

Figure 1

In this plot, you can clearly see the three different "pieces" of the function. (The vertical segment in the graph is an artifact of *Maple*'s plotting routine, and is not part of the function.)

EXERCISE. Write down an expression for the following function in terms of the **Heaviside** function, and then plot it:

$$f(t) = \begin{cases} 0, & t < 1, \\ \cos t, & 1 \le t < 2, \\ \sin t, & t \ge 2. \end{cases}$$

Another important discontinuous function is the *Dirac delta function*, usually denoted $\delta(t)$ or $\delta_0(t)$, and sometimes called the *unit impulse function*. To understand this function, recall that the total impulse imparted by a varying force $F(t)$ is the integral of $F(t)$. If we consider a constant force of total impulse one and make this force act over a smaller and smaller time interval around 0, then in the limit we obtain the Dirac delta function. Thus, the delta function represents an idealized force of total impulse 1 concentrated at the instant $t = 0$. The delta function belongs to a class of mathematical objects called *generalized functions*. (It is not a true function since its value at 0 would be infinite.) The magnitude and timing of the delta function can be changed by multiplying the delta function by a constant and translating it. Thus, the function $10\delta(t-8)$ represents a force of total impulse 10 concentrated at the instant $t = 8$. The function $\delta(t-c)$ is sometimes denoted $\delta_c(t)$. The Dirac delta function is implemented in *Maple*, under the name **Dirac**.

EXERCISE. The fact that **Dirac** is not a true function causes unusual behavior in *Maple*. Plot the Dirac function on the interval $[-1, 1]$ (you won't see anything). Evaluate the function at 0 and 1. Integrate the function over the interval $[-1, 1]$.

Both the unit step function and the Dirac delta function have Laplace Transforms. For example,

$$\mathcal{L}(u_c)(s) = \int_0^\infty u_c(t) e^{-st}\, dt = \int_c^\infty e^{-st}\, dt$$

$$= \frac{e^{-cs}}{s}, \quad \text{for } s > 0.$$

To compute the Laplace Transform of $\delta(t)$, we must write $\delta(t)$ as a limit of *bona fide* functions. The easiest way to do this is to use step functions. We can write a function of total integral one, concentrated on the interval $[-\epsilon, \epsilon]$, as

$$\frac{1}{2\epsilon}(u_{-\epsilon}(t) - u_\epsilon(t)).$$

Thus

$$\delta(t) = \lim_{\epsilon \to 0+} \frac{1}{2\epsilon}(u_{-\epsilon}(t) - u_\epsilon(t)),$$

and

$$\mathcal{L}(\delta_c)(s) = \lim_{\epsilon \to 0+} \frac{1}{2\epsilon}(\mathcal{L}(u_{c-\epsilon})(s) - \mathcal{L}(u_{c+\epsilon})(s))$$

$$= \lim_{\epsilon \to 0+} \frac{1}{2\epsilon}\left(\frac{e^{(\epsilon-c)s}}{s} - \frac{e^{-(\epsilon+c)s}}{s}\right)$$

$$= \lim_{\epsilon \to 0+} (e^{-cs})\frac{e^{\epsilon s} - e^{-\epsilon s}}{2\epsilon s}$$

$$= (e^{-cs}) \lim_{\epsilon \to 0+} \frac{s(e^{\epsilon s} + e^{-\epsilon s})}{2s}$$

$$= (e^{-cs}) \lim_{\epsilon \to 0+} \cosh(\epsilon s)$$

$$= e^{-cs}.$$

(We used L'Hôpital's rule in passing from the third to the fourth line.) Thus $\mathcal{L}(\delta_c)(s) = e^{-cs}$. In particular, setting $c = 0$, we get that $\mathcal{L}(\delta)(s) = 1$.

Additional properties of the Laplace Transform are explored in Problems 10 and 11 of Problem Set E.

Differential Equations with Discontinuous Forcing Functions

Consider an inhomogeneous, second order linear equation with constant coefficients:

$$ay''(t) + by'(t) + cy(t) = g(t).$$

The function $g(t)$ is called the forcing function of the differential equation, because in many physical models $g(t)$ corresponds to the influence of an external force. If $g(t)$ is piecewise continuous, or involves the delta function, then we can solve the equation by the method of Laplace Transforms.

EXAMPLE. Consider the initial value problem:

$$y''(t) + 3y'(t) + y(t) = g(t), \quad y(0) = 1, \quad y'(0) = 1,$$

where

$$g(t) = \begin{cases} 0, & t < 0, \\ 1, & 0 \le t < 1, \\ -1, & 1 \le t < 2, \\ 0, & t \ge 2. \end{cases}$$

We can solve this equation in *Maple* with the following sequence of commands.

```
g := t -> Heaviside(t) - 2*Heaviside(t - 1) + Heaviside(t - 2);
eqn3 := diff(y(t), t$2) + 3*diff(y(t), t) + y(t) = g(t);
sol3 := dsolve({eqn3, y(0) = 1, D(y)(0) = 1}, y(t), method = laplace);
```

The solution is too complicated to reproduce here (but you should compute it using *Maple*). Here is a graph of the solution together with the forcing function, produced with the *Maple* command **plot({rhs(sol3), g(t)}, t = 0..5);**

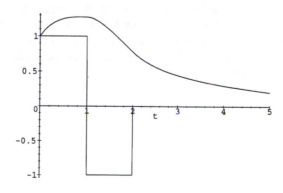

Figure 2

EXAMPLE. We can also use *Maple* to solve nonhomogeneous equations involving the delta function. Consider the initial value problem

$$y''(t) + y'(t) + y(t) = \delta_1(t), \quad y(0) = 0, y'(0) = 0.$$

We can solve this equation with the following sequence of commands.

eqn4 := diff(y(t), t$2) + diff(y(t), t) + y(t) = Dirac(t − 1);
sol4 := dsolve({eqn4, y(0) = 0, D(y)(0) = 0}, y(t), method = laplace);

The solution is

$$y(t) = \frac{2\sqrt{3}\,\text{Heaviside}(t-1)e^{-(t-1)/2}\sin(\sqrt{3}(t-1)/2)}{3}.$$

Here is a graph of the solution.

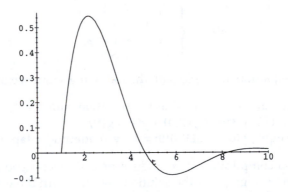

Figure 3

In this graph, you can clearly see the effect of the unit impulse at time $t = 1$. Until $t = 1$, the solution is 0. At time $t = 1$, the unit impulse puts the solution into a nonzero state. After $t = 1$, the impulse function has no further effect. Since the roots of the characteristic polynomial of the homogeneous equation have negative real part, the solution decays exponentially to 0.

Problem Set E

Series Solutions
and Laplace Transforms

1. Consider Airy's equation $y'' - xy = 0$.

   (a) Compute the terms of degree 10 or less in the Taylor series expansion of the solution $y(x)$ to Airy's equation $y'' - xy = 0$ that satisfies $y(0) = 1$, $y'(0) = 0$.

   (b) On the interval $(0, 6)$, graph the Taylor polynomial you obtained, together with the exact solution of the initial value problem.

   (c) Airy's equation can be written as $y'' = xy$. If $x > 1$, then $y'' = xy > y$, so the solutions of Airy's equation grow at least as quickly as the solutions of $y'' = y$. Analyze the latter equation, and predict how well Taylor polynomials will approximate exact solutions of Airy's equation. Relate your analysis to the graph in part (b).

2. Compute the terms of degree 10 or less in the Taylor series expansion of the solution to Problem 23 in Section 5.2 of Boyce & DiPrima:

$$y'' - xy' - y = 0, \quad y(0) = 1, \quad y'(0) = 0.$$

   (a) Graph the Taylor polynomial of degree 10.

   (b) What is the long-term behavior of the Taylor polynomial (as $x \to \pm\infty$)?

   (c) Is the Taylor polynomial an even function? odd? neither?

   (d) Explain purely from the IVP itself why each of your conclusions holds for the actual solution. (*Hints*: In (b), you will have to do a little qualitative analysis—approximate x by a constant; in (c), consider $y_1(x) = y(-x)$.)

3. Compute the terms of degree 10 or less in the Taylor series expansion of the solution to Problem 18 in Section 5.2 of Boyce & DiPrima:

$$(1 - x)y'' + xy' - y = 0, \quad y(0) = -3, \quad y'(0) = 2.$$

The theory suggests the true solution may have a singularity at $x = 1$. Graph the Taylor polynomial you obtained. Do you see any signs of the singularity? Compute the exact solution using **dsolve**. Does the solution have a singularity?

149

4. Use **dsolve** to solve the following Euler equations in Section 5.5 of Boyce & DiPrima (Probs. 3, 6, 8, 13):

(a) $x^2 y'' - 3xy' + 4y = 0$

(b) $(x-1)^2 y'' + 8(x-1)y' + 12y = 0$

(c) $2x^2 y'' - 4xy' + 6y = 0$

(d) $2x^2 y'' + xy' - 3y = 0, \quad y(1) = 1, \quad y'(1) = 4.$

5. Compute the terms of degree 10 or less in the Frobenius series expansion of the solution to the following problems in Section 5.6 of Boyce & DiPrima (Probs. 2, 5, 8):

(a) $x^2 y'' + xy' + (x^2 - \frac{1}{9})y = 0$

(b) $3x^2 y'' + 2xy' + x^2 y = 0$

(c) $2x^2 y'' + 3xy' + (2x^2 - 1)y = 0.$

6. This problem concerns Bessel's equation, which arises in many physical problems with circular symmetry, such as the study of water waves in a circular pond, vibrations of a circular drum, or the diffraction of light by the circular aperture of a telescope. The Bessel function $J_n(x)$ (where the parameter n is an arbitrary integer) is defined to be the coefficient of t^n in the power series expansion of the function

$$\exp\left(\frac{1}{2}x\left(t - \frac{1}{t}\right)\right).$$

(By thinking about what happens to the power series when t and $1/t$ are switched, you can see that $J_{-n}(x) = J_n(x)$ for n even and $J_{-n}(x) = -J_n(x)$ for n odd.) $J_n(x)$ is also a solution to *Bessel's equation of order n*,

$$x^2 y'' + xy' + (x^2 - n^2)y = 0.$$

The Taylor series expansion for J_0 is derived in Section 5.8 of Boyce & DiPrima:

$$J_0(x) = \sum_{j=0}^{\infty}(-1)^j \frac{1}{2^{2j}} \frac{x^{2j}}{(j!)^2}.$$

(a) *Maple* has a built-in function for $J_n(x)$, called **BesselJ(n, x)**. Use it to plot $J_0(x)$ for $0 \leq x \leq 10$. (You don't need to graph the function for negative x since J_0 is an even function.) On the same axes, draw the graphs of the 4th order, 10th order, and 20th order Taylor polynomials of J_0. (Use the **series** command in *Maple* (type **?series** for a description) to compute the Taylor expansion of a function to a prescibed order.) How well do the Taylor polynomials approximate the function?

(b) An interesting equality satisfied by the Bessel functions is

$$J_n'(x) = \frac{1}{2}\left(J_{n-1}(x) - J_{n+1}(x)\right).$$

In particular,

$$J_0'(x) = \frac{1}{2}\left(J_{-1}(x) - J_1(x)\right) = \frac{1}{2}\left(-J_1(x) - J_1(x)\right) = -J_1(x).$$

Thus, the maxima and minima of J_0 occur at the zeros of J_1. Using **fsolve**, solve the equations $J_0(x) = 0$ and $J_1(x) = 0$ numerically to find the first five zeros and the first five relative extreme points of J_0, starting at $x = 0$ (you will need lots of different starting values—use your graph for guidance). Compute the differences between successive zeros. Can you make a guess about the periods of the oscillations as $x \to \infty$? (If you did Problem 15 in Problem Set D, you know exactly how the period behaves as $x \to \infty$.)

(c) From the general theory, one knows that Bessel's equation of order n has another linearly independent solution $Y_n(x)$, with a logarithmic singularity as $x \to 0^+$. *Maple* also has a built-in function, **BesselY(n, x)**, for computing $Y_n(x)$. Graph the function $Y_0(x)$ on the interval $0 < x \leq 10$. Then compute $\lim_{x \to 0^+} Y_0(x)/\ln x$. Graph the function

$$Y_0(x) - c \ln x$$

(on the same interval), where c is the value of the limit you obtained, and observe that it is continuous in a neighborhood of $x = 0$. Thus, the "singular part" of $Y_0(x)$ behaves like $c \ln x$.

7. The first order homogeneous linear differential equation

$$x^2 y' + y = 0$$

has an irregular singular point at $x = 0$ (why?); however, it happens to be an equation that is easy to solve exactly. Find the general solution of the equation for $x > 0$ and for $x < 0$. How do the solutions behave as $x \to 0^+$? as $x \to 0^-$? Use *Maple* to graph a solution for $0 < x < 2$ and a solution for $-2 < x < 0$.

Now find (for $x > 0$ and for $x < 0$) the general solution of the Euler equation

$$xy' + ry = 0,$$

which has a regular singular point at $x = 0$ if $r \neq 0$. How do the solutions behave as $x \to 0^+$? as $x \to 0^-$? Use *Maple* to graph solutions to the right and left of 0 for each of the cases $r = 0.5$, $r = 1$, $r = -0.5$, $r = -1$. What difference do you notice between the behavior of solutions near a regular singular point and the behavior near an irregular singular point?

8. Consider the Legendre equation

$$(1 - x^2)y'' - 2xy' + \alpha(\alpha + 1)y = 0.$$

We assume that $|x| < 1$, because $x = \pm 1$ are regular singular points. We also assume that $\alpha > -1$.

(a) Find series approximations of the two solutions of the Legendre equation with initial conditions $y(0) = 1$, $y'(0) = 0$, and $y(0) = 0$, $y'(0) = 1$. Is this a fundamental set of solutions?

(b) If $\alpha = n$ is a nonnegative integer, the Legendre equation has a polynomial solution of order n. The *Legendre polynomial* $P_n(x)$ is defined to be the unique polynomial solution of the Legendre equation (with $\alpha = n$) such that $P_n(1) = 1$. Use *Maple* to compute the first six Legendre polynomials $P_0(x), \ldots, P_5(x)$. You can do this by using **dsolve** (without the series option and without initial conditions) to solve the Legendre equation for $n = 0, \ldots, 5$; then by inspecting each solution, choose values for the undetermined constants to get polynomial solutions with $P_n(1) = 1$.

(c) Graph $P_0(x), \ldots, P_5(x)$ on the interval $[-1, 1]$. Use different line styles so that you can distinguish the different curves.

(d) In the graph in part (c), you can see that as n increases, the Legendre polynomials are more and more oscillatory on the interval $[-1, 1]$. For x close to zero, the Legendre equation is approximately $y'' + n(n+1)y = 0$. By solving this equation, explain why solutions of the Legendre equation display increasing oscillation (on the interval $[-1, 1]$) as $n \to \infty$.

9. Nonlinear differential equations can sometimes be solved by series methods. Consider, for example, the equation

$$y' = y^2 + x, \qquad (i)$$

mentioned in Chapter 7 as an example of an equation that cannot be solved in terms of elementary functions. Consider this equation with the initial condition $y(0) = 0$, and let's look for a solution of the form

$$y(x) = \sum_{i=1}^{\infty} a_i x^i. \qquad (ii)$$

(We know there is no constant term in the Taylor series expansion of y at $x = 0$ because of the initial condition.) Substituting (ii) into (i) and expanding the right-hand side gives a sequence of simultaneous quadratic equations for the coefficients a_i that become increasingly complicated as $i \to \infty$. However, we can solve inductively for the coefficients, by first equating the terms in x, then the terms in x^2, etc. *Maple* can carry out this task.

(a) Explain why a_1 must be 0.

(b) Now use **dsolve** to compute a Taylor series expansion of the solution to (i). Make sure you go far enough to get the 11th degree monomial. What pattern do you see?

(c) Plot the 11th degree Taylor polynomial for y, together with a numerical solution of (i) obtained using **dsolve(..., numeric)**, on the interval $-1 < x < 1$. How close are the graphs? Now note that $y(x) > x^2/2$ for $0 < x < 1$, as you can see, for example, from the series. Thus, $y(1) > 0.5$. But for $x > 1$, the right-hand side of (i) is greater than $y(x)^2 + 1$, so the solution grows faster than the solution to the IVP $y' = y^2 + 1$, $y(1) = 0.5$, which is of the form $y(x) = \tan(x - C)$, where $C \approx 0.54$. Since the tangent function blows up when its argument reaches $\pi/2$, this analysis shows that the solution of (i) must become infinite somewhere between $x = 1$ and $x = 2.2$. Is there any way of seeing this from the series solution? Plot the numerical solution to locate the point P where the solution blows up. Graph the two approximate solutions (that is, the series solution out to degree 11 and the numerical solution on the interval $[1, 2]$). What do you observe?

10. Suppose $f(t)$ and $g(t)$ are functions with Laplace Transforms $\mathcal{L}(f) = F$ and $\mathcal{L}(g) = G$. Find out which of the following identities *Maple* knows about. (For some of these, you may have to enter the command **assume(c > 0)** to tell *Maple* that the constant is positive. Once you have entered this command, *Maple* will print out the constant c with a tilde after it, as $c\tilde{}$, to indicate that c is a *restricted* variable. To clear c, just type **c := 'c'.**)

(a) $\mathcal{L}(af + bg) = a\mathcal{L}(f) + b\mathcal{L}(g)$, where a and b are real numbers.

(b) $\mathcal{L}(u_c(t)f(t - c)) = e^{-cs}F(s)$

(c) $\mathcal{L}(e^{ct}f(t)) = F(s - c)$

(d) $\mathcal{L}(f') = sF(s) - f(0)$

(e) $\mathcal{L}(\int_0^t f(t - u)g(u)\,du) = F(s)G(s)$.

The function $t \to \int_0^t f(t - u)g(u)\,du$ is called the *convolution* of f and g, and is written $f * g$. Thus, the Laplace Transform changes convolution into ordinary multiplication.

11. For each of the following functions, compute the Laplace Transform. Then plot the function and its Laplace Transform over the interval $[0, 10]$. You may want to restrict the vertical range for some of the graphs, and to use different plot styles for the function and its Laplace Transform.

(a) $\sin t$

(b) $\exp t$

(c) $\dfrac{1}{t + 1}$

(d) $\cos t$

(e) $t \cos t$

(f) $u_1(t) \sin t$

On the basis of these graphs, what general conclusions can you draw about the Laplace Transform of a function? In particular, what can you say about the growth or decay of the Laplace Transform of a function? Can you justify your conclusion by looking at the integral formula for the Laplace Transform? (*Hint:* Look at the derivation of $\mathcal{L}(e^{at})$ on the second page of Chapter 11.)

12. This problem illustrates how the choice of method can dramatically affect the time it takes the computer to solve a differential equation. Consider the IVP

$$y'' + y' + y = t^2 e^{-t} \cos t, \qquad y(0) = 1, \qquad y'(0) = 0.$$

This problem could be solved by the method of undetermined coefficients, variation of parameters, or Laplace Transforms.

(a) Try to solve the IVP using **dsolve**. Execute the command **time()** before and after you execute **dsolve** to measure the amount of time *Maple* requires to solve the problem. (*Maple* uses the method of variation of parameters.)

(b) Now use **dsolve** with the option **method = laplace**. Once again, execute **time()** before and after you execute **dsolve**. How does the elapsed time compare to (a)?

(c) Define a new function $g(t)$ to be the output from part (b). Verify that g is a solution to the differential equation. (Don't forget **simplify**.)

(d) Now repeat part (b) with the symbolic initial conditions $y(0) = y0$, $y'(0) = y1$ (you need not keep track of time). Can *Maple* compute Laplace Transform solutions with symbolic initial conditions?

13. Consider the following IVP:

$$y'' + y' + \frac{5}{4}y = g(t), \quad y(0) = 0, \quad y'(0) = 0, \quad g(t) = \begin{cases} \sin t, & \text{if } 0 \le t < \pi \\ 0, & \text{if } t \ge \pi. \end{cases}$$

(a) Solve this equation using **dsolve** (without **method = laplace**). Plot the solution on the interval $[0, 15]$.

(b) Check to see whether the function produced by **dsolve** is actually a solution by substituting it back into the differential equation. The expression will be quite complicated. Evaluate it numerically at $t = 0, 0.5,$ and 1, to see if both sides of the equation are the same.

(c) Now solve the equation using **dsolve** with **method = laplace**. Plot the solution on $[0, 15]$. Check to see whether this solution is correct by substituting into the original equation.

(d) Find the general solution of the associated homogeneous equation. What is the asymptotic behavior of those solutions as $t \to \infty$? Knowing that the forcing term of the inhomogeneous equation is zero after $t = \pi$, what can you say about the long-term behavior of the solution of the inhomogeneous equation?

14. Use the Laplace Transform method to solve the following IVPs. Then graph the solution together with the function on the right hand side of the differential equation (the forcing function) on the interval $[0, 15]$.

   (a) $y'' + 2y' + 2y = h(t)$, $\quad y(0) = 0$, $\quad y'(0) = 1$, $\quad h(t) = \begin{cases} 1, & \text{if } \pi \le t < 2\pi \\ 0, & \text{otherwise} \end{cases}$

   (b) $y'' + 3y' + 2y = u_2(t)$, $\quad y(0) = 0$, $\quad y'(0) = 1$

   (c) $y'' + y = \delta(t - \pi) \cos t$, $\quad y(0) = 0$, $\quad y'(0) = 1$

   (d) $y'' + 2y' + 2y = \cos t + \delta(t - \pi/2)$, $\quad y(0) = 0$, $\quad y'(0) = 0$.

15. Use the Laplace Transform method to solve the following IVPs. Then graph the solution on an appropriate interval $(t > 0)$.

   (a) $y'''(t) - y''(t) - y'(t) + y(t) = \delta(t - 1)$, $\quad y(0) = y''(0) = 0$, $\quad y'(0) = 1$

   (b) $y^{(4)}(t) + 2y''(t) + y(t) = \sin t$, $\quad y(0) = 1$, $\quad y'(0) = y''(0) = y'''(0) = 0$

   (c) $y^{(4)}(t) + 5y''(t) + 4y(t) = u_1(t) \sin(3t)$, $\quad y(0) = y'(0) = y''(0) = y'''(0) = 0$

   (d) $y'''(t) + y''(t) + y'(t) = \delta(t - 1)$, $\quad y(0) = y''(0) = 0$, $\quad y'(0) = 1$.

   Which of these equations has resonance-type behavior?

16. In this problem we investigate the effect of a periodic discontinuous forcing function (a square wave) on a second order linear equation with constant coefficients. Consider the initial value problem

$$y'' + y = h(t), \quad y(0) = 0, \, y'(0) = 1. \tag{i}$$

The associated homogeneous equation has natural period 2π; the general solution of the homogeneous equation is

$$y(t) = A \cos t + B \sin t.$$

Recall that the phenomenon of resonance occurs when the forcing function $h(t)$ is a linear combination of $\sin t$ and $\cos t$. Does resonance occur when the forcing function is periodic of period 2π but discontinuous?

(a) Using step functions, define a *Maple* function $h(t)$ on the interval $[0, 10\pi]$ whose value is $+1$ on $[0, \pi)$, -1 on $[\pi, 2\pi)$, $+1$ on $[2\pi, 3\pi)$, and so on. Plot the function on the interval $[0, 30]$. It should have the appearance of a square wave.

(b) Use **dsolve** with **method = laplace** to solve equation (i) with the function $h(t)$ defined in part (a). Plot the solution together with $h(t)$ on the interval $[0, 30]$. Do you see resonance? Compute and plot a numerical solution to confirm your answer.

(c) In part (a), we constructed a forcing function $h(t)$ with period 2π. The function $h(t/2)$ has period 4π. Repeat part (b) using the forcing function $h(t/2)$. Do you see resonance?

(d) Repeat part (b) using the forcing function $h(2t)$ (this time just plot from 0 to 15). Do you see resonance? What is the period of $h(2t)$?

(e) What can you conclude about the resonance effect for discontinuous forcing functions? Would you expect resonance to occur in equation (i) for *any* forcing function of period 2π? (*Hint:* The function $h(2t)$ has period π as well as period 2π.) Can you venture a guess about when resonance occurs for discontinuous periodic forcing functions? You might try some other periodic forcing functions to check your guess.

17. Do Problem 16 with a sawtooth wave instead of a square wave. More specifically, replace part (a) of Problem 16 with the following, and then do parts (b)–(e) of Problem 16.

(a) Using step functions, define a *Maple* function $h(t)$ on the interval $[0, 10\pi]$, with $h(t) = t$ on $[0, \pi)$, $h(t) = t - \pi$ on $[\pi, 2\pi)$, $h(t) = t - 2\pi$ on $[2\pi, 3\pi)$, and so on. Plot the function on the interval $[0, 30]$. It should have the appearance of a sawtooth wave.

18. This problem is based on Problem 35 in Section 6.2 of Boyce & DiPrima. Consider Bessel's equation
$$ty'' + y' + ty = 0.$$
We will use the Laplace Transform to find some leading terms of the power series expansion of the Bessel function of order zero that is continuous at the origin. Let $y(t)$ be such a solution of the equation, and suppose $Y(s)$ is its transform.

(a) By taking the Laplace Transform of the differential equation, show that $Y(s)$ satisfies the equation
$$(1 + s^2)Y'(s) + sY(s) = 0.$$

(b) Solve the equation in part (a) using **dsolve**. (Use $Y(0) = 1$.)

(c) Use the **series** command to expand the solution in part (b) in powers of $1/s$ out to terms of degree 12. (*Hint:* Expand around $s = \infty$.)

(d) Apply the inverse Laplace Transform to part (c) to obtain the power series expansion of y up to degree 10.

Chapter 12

Higher Order Equations and Systems of First Order Equations

First and second order differential equations arise naturally in many applications. For example, first order differential equations occur in models of population growth and radioactive decay, and second order equations in the study of the motion of a falling body or the motion of a pendulum. There are several techniques for solving special classes of first and second order equations. These techniques produce symbolic solutions, expressed by a formula. For equations that cannot be solved by any of these techniques, we use numerical, geometric, or qualitative methods to investigate solutions.

Equations of higher order, and systems of first order equations, also arise naturally. For example, third order equations come up in fluid dynamics; fourth order equations, in elasticity; and systems of first order equations, in the study of spring–mass systems with several springs and masses. For general higher order equations and systems, there are hardly any techniques for obtaining explicit formula solutions, and numerical, geometric, or qualitative techniques must be used. Nevertheless, for the special class of constant coefficient linear equations and first order linear systems, there are techniques for producing explicit solutions.

In this chapter, we consider higher order equations and first order systems. We outline the basic theory in the linear case. We show how to solve linear first order systems, first by using *Maple* to compute eigenpairs of the coefficient matrix, and then by using **dsolve**. In addition, we discuss the the plotting of phase portraits using *Maple*.

Higher Order Linear Equations

Consider the linear equation

$$y^{(n)} + p_1(x)y^{(n-1)} + \cdots + p_n(x)y = g(x), \tag{1}$$

and assume the coefficient functions are continuous on the interval $a < x < b$. Then the following are true:

- If $y_1(x), y_2(x), \ldots, y_n(x)$ are n linearly independent solutions of the corresponding homogeneous equation,

$$y^{(n)} + p_1(x)y^{(n-1)} + \cdots + p_n(x)y = 0, \tag{2}$$

  and if y_p is any particular solution of the inhomogeneous equation (1), then the general solutions of the equations (1), (2) are, respectively,

$$y = \sum_{i=1}^{n} C_i y_i(x) + y_p(x)$$

  and

$$y = \sum_{i=1}^{n} C_i y_i(x).$$

- Given a point $x_0 \in (a, b)$ and initial values $y(x_0) = y_0$, $y'(x_0) = y_0', \ldots,$ $y^{(n-1)}(x_0) = y_0^{(n-1)}$, there is a unique solution to equation (1) that satisfies the initial data; the solution has n continuous derivatives and exists on the entire interval (a, b).

- A set of n solutions $y_1(x), y_2(x), \ldots, y_n(x)$ of (2) is linearly independent if and only if the Wronskian

$$W(y_1, y_2, \ldots, y_n) = \det \begin{pmatrix} y_1 & y_2 & \cdots & y_n \\ y_1' & y_2' & \cdots & y_n' \\ \vdots & \vdots & \ddots & \vdots \\ y_1^{(n-1)} & y_2^{(n-1)} & \cdots & y_n^{(n-1)} \end{pmatrix}$$

  is nonzero at some point in (a, b); the collection $y_1, \ldots, y_n$ is called a *fundamental set* of solutions of (2).

We see that the general theory of higher order linear equations parallels that of first and second order equations. We note, however, that solutions may be difficult to construct. In practice, we can construct fundamental sets of solutions only for constant coefficient and Euler equations. In these cases, solutions can be constructed by considering trial solutions e^{rx} and x^r, as in the second order case. There is a difficulty even for these equations if n is large, since it is difficult to compute the roots of the characteristic polynomial. We can, however, use *Maple* to find good approximations to the roots.

The **dsolve** command can solve many homogeneous constant coefficient and Euler equations. You can confirm this by solving several linear equations of higher order from your text.

Systems of First Order Equations

Consider the general system of first order equations,

$$x_1' = F_1(t, x_1, x_2, \ldots, x_n)$$
$$x_2' = F_2(t, x_1, x_2, \ldots, x_n)$$
$$\vdots \tag{3}$$
$$x_n' = F_n(t, x_1, x_2, \ldots, x_n).$$

Such systems arise in the study of spring–mass systems with many springs and masses, and in many other applications. Systems of first order equations also arise in the study of higher order equations. A single higher order differential equation

$$y^{(n)} = F(t, y, y', \ldots, y^{(n-1)}),$$

can be converted into a system of first order equations, simply by setting

$$x_1 = y, x_2 = y', \ldots, x_n = y^{(n-1)}.$$

The resulting system is

$$x_1' = x_2$$
$$x_2' = x_3$$
$$\vdots$$
$$x_{n-1}' = x_n$$
$$x_n' = F(t, x_1, x_2, \ldots, x_n).$$

It is not surprising that the problem of analyzing the general system (3) is quite formidable. Generally, the techniques we can bring to bear will be numerical, geometric, or qualitative. We shall discuss the application of these techniques to nonlinear equations in Chapter 13. For now, we merely observe that we can expect to make some progress toward a formula solution for *linear systems*. So, consider a linear system

$$x_1' = a_{11}(t)x_1 + \ldots + a_{1n}(t)x_n + b_1(t)$$
$$x_2' = a_{21}(t)x_1 + \ldots + a_{2n}(t)x_n + b_2(t)$$
$$\vdots \quad \vdots$$
$$x_n' = a_{n1}(t)x_1 + \ldots + a_{nn}(t)x_n + b_n(t).$$

We can write this system in matrix notation:

$$\mathbf{X}' = \mathbf{A}(t)\mathbf{X} + \mathbf{B}(t),$$

where

$$
\mathbf{X} = \begin{pmatrix} x_1(t) \\ x_2(t) \\ \vdots \\ x_n(t) \end{pmatrix}, \qquad
\mathbf{B} = \begin{pmatrix} b_1(t) \\ b_2(t) \\ \vdots \\ b_n(t) \end{pmatrix},
$$

and

$$
\mathbf{A} = \begin{pmatrix}
a_{11}(t) & a_{12}(t) & \cdots & a_{1n}(t) \\
\vdots & \vdots & \ddots & \vdots \\
a_{n1}(t) & a_{n2}(t) & \cdots & a_{nn}(t)
\end{pmatrix}.
$$

There is a general theory for linear systems, which is completely parallel to the theory for linear differential equations. If $\mathbf{X}^{(1)}, \mathbf{X}^{(2)}, \ldots, \mathbf{X}^{(n)}$ is a family of linearly independent solutions, *i.e.*, a *fundamental set* of solutions, of the homogeneous equation (with $\mathbf{B} = 0$), and if $\mathbf{X}_p$ is a particular solution of the inhomogeneous equation, then the general solutions of these equations are, respectively,

$$
\mathbf{X} = \sum_{i=1}^{n} C_i \mathbf{X}^{(i)}
$$

and

$$
\mathbf{X} = \sum_{i=1}^{n} C_i \mathbf{X}^{(i)} + \mathbf{X}_p.
$$

The test for linear independence of n solutions is

$$
\det\left(\mathbf{X}^{(1)} \ldots \mathbf{X}^{(n)}\right) \neq 0.
$$

For linear systems, the constant coefficient case is the easiest to handle, just as it was for single linear equations. Consider the homogeneous linear system

$$
\mathbf{X}' = \mathbf{A}\mathbf{X}, \tag{4}
$$

where

$$
\mathbf{A} = \begin{pmatrix}
a_{11} & a_{12} & \cdots & a_{1n} \\
\vdots & \vdots & \ddots & \vdots \\
a_{n1} & a_{n2} & \cdots & a_{nn}
\end{pmatrix}
$$

is an $n \times n$ matrix of constants. We seek solutions of the form $\mathbf{X}(t) = \xi e^{rt}$; note the similarity to the case of a single linear equation. We find that $\mathbf{X}(t)$ is a nonzero solution of (4) if and only if

$$
\mathbf{A}\xi = r\xi,
$$

i.e., if and only if r is an *eigenvalue* of $\mathbf{A}$ and ξ is a corresponding *eigenvector*. We call the pair r, ξ an *eigenpair*. In order to explain how to construct a fundamental set of solutions, we have to consider three separate cases.

Case I. **A** has distinct real eigenvalues.

Let $r_1, r_2, \ldots, r_n$ be the distinct eigenvalues of **A**, and let $\xi^{(1)}, \xi^{(2)}, \ldots, \xi^{(n)}$ be corresponding eigenvectors. Then

$$\mathbf{X}^{(1)}(t) = \xi^{(1)} e^{r_1 t}, \ldots, \mathbf{X}^{(n)}(t) = \xi^{(n)} e^{r_n t}$$

is a fundamental set of solutions for (4), and

$$\mathbf{X}(t) = c_1 \xi^{(1)} e^{r_1 t} + \cdots + c_n \xi^{(n)} e^{r_n t}$$

is the general solution.

Case II. **A** has distinct eigenvalues, some of which are complex.

Since **A** is real, if r, ξ is a complex eigenpair, then the complex conjugate $\bar{r}, \bar{\xi}$ is also an eigenpair. Thus, the corresponding solutions

$$\mathbf{X}^{(1)}(t) = \xi e^{rt}, \quad \mathbf{X}^{(2)}(t) = \bar{\xi} e^{\bar{r} t}$$

are conjugate. Therefore, we can find two real solutions of (4) corresponding to the pair $r, \bar{r}$ by taking the real and imaginary parts of $\mathbf{X}^{(1)}(t)$ or $\mathbf{X}^{(2)}(t)$. Writing $\xi = \mathbf{a} + i\mathbf{b}$, where **a** and **b** are real, and $r = \lambda + i\mu$, where λ and μ are real, we have

$$\mathbf{X}^{(1)}(t) = (\mathbf{a} + i\mathbf{b}) e^{(\lambda + i\mu)t}$$

$$= (\mathbf{a} + i\mathbf{b}) e^{\lambda t} (\cos \mu t + i \sin \mu t)$$

$$= e^{\lambda t} (\mathbf{a} \cos \mu t - \mathbf{b} \sin \mu t) + i e^{\lambda t} (\mathbf{a} \sin \mu t + \mathbf{b} \cos \mu t).$$

Thus, the vector functions

$$\mathbf{u}(t) = e^{\lambda t} (\mathbf{a} \cos \mu t - \mathbf{b} \sin \mu t)$$

$$\mathbf{v}(t) = e^{\lambda t} (\mathbf{a} \sin \mu t + \mathbf{b} \cos \mu t)$$

are real solutions to (4).

To keep the discussion simple, assume that $r_1 = \lambda + i\mu$, $r_2 = \lambda - i\mu$ are complex, and that $r_3, \ldots, r_n$ are real and distinct. Let the corresponding eigenvectors be $\xi^{(1)} = \mathbf{a} + i\mathbf{b}$, $\xi^{(2)} = \mathbf{a} - i\mathbf{b}$, $\xi^{(3)}, \ldots, \xi^{(n)}$. Then

$$\mathbf{u}(t), \mathbf{v}(t), \xi^{(3)} e^{r_3 t}, \ldots, \xi^{(n)} e^{r_n t}$$

is a fundamental set of solutions for (4). The general situation should now be clear.

Case III. **A** has repeated eigenvalues.

We restrict our discussion to the case $n = 2$. Suppose $r = \rho$ is an eigenvalue of **A** of multiplicity 2, meaning that ρ is a double root of the characteristic polynomial $\det(\mathbf{A} - r\mathbf{I})$. There are still two possibilities. If we can find two linearly independent eigenvectors $\xi^{(1)}$ and $\xi^{(2)}$ corresponding to ρ, then the solutions $\mathbf{X}^{(1)}(t) = \xi^{(1)} e^{\rho t}$ and $\mathbf{X}^{(2)}(t) = \xi^{(2)} e^{\rho t}$ form a fundamental set. The other possibility is that there is only one (linearly independent) eigenvector ξ with eigenvalue ρ. Then one solution of (4) is given by

$$\mathbf{X}^{(1)}(t) = \xi e^{\rho t},$$

and a second by

$$\mathbf{X}^{(2)}(t) = \boldsymbol{\xi} t e^{\rho t} + \boldsymbol{\eta} e^{\rho t},$$

where η satisfies

$$(\mathbf{A} - r\mathbf{I})\eta = \boldsymbol{\xi}.$$

The vector η is called a *generalized eigenvector* of $\mathbf{A}$ for the eigenvalue ρ.

Using *Maple* to Find Eigenpairs

We now show how to use *Maple* to find the eigenpairs of a matrix, and thus to find fundamental sets of solutions and the general solution of a system of linear differential equations with constant coefficients. In order to manipulate matrices with *Maple*, you must first load the **linalg** package.

EXAMPLE 1. Consider the system

$$\mathbf{x}' = \begin{pmatrix} 3 & -2 & 0 \\ 2 & -2 & 0 \\ 0 & 1 & 1 \end{pmatrix} \mathbf{x}.$$

We enter the coefficient matrix in *Maple* as a list of lists:

> **with(linalg):**
> **A := [[3, −2, 0], [2, −2, 0], [0, 1, 1]];**

We can display $\mathbf{A}$ in standard matrix form using the command **matrix(A)**, which produces

$$\begin{bmatrix} 3 & -2 & 0 \\ 2 & -2 & 0 \\ 0 & 1 & 1 \end{bmatrix}$$

We find the eigenpairs of $\mathbf{A}$ with the command **eigenvects(A)**, which produces

$$[-1, 1, \{[-1, -2, 1]\}], [1, 1, \{[0, 0, 1]\}], [2, 1, \{[2, 1, 1]\}]$$

The output of this command is a sequence; each entry of the sequence is a triple $[\rho, m, [eigenvectors]]$ consisting of an eigenvalue ρ, the number m of times it occurs, and a set of independent eigenvectors associated to it. In this example, the eigenpairs are

$$r_1 = -1, \; \boldsymbol{\xi}^{(1)} = \begin{pmatrix} -1 \\ -2 \\ 1 \end{pmatrix}$$

$$r_2 = 1, \; \boldsymbol{\xi}^{(2)} = \begin{pmatrix} 0 \\ 0 \\ 1 \end{pmatrix}$$

$$r_3 = 2, \; \boldsymbol{\xi}^{(3)} = \begin{pmatrix} 2 \\ 1 \\ 1 \end{pmatrix}.$$

Thus, a fundamental set of solutions is

$$\mathbf{X}^{(1)}(t) = \begin{pmatrix} -1 \\ -2 \\ 1 \end{pmatrix} e^{-t}, \quad \mathbf{X}^{(2)}(t) = \begin{pmatrix} 0 \\ 0 \\ 1 \end{pmatrix} e^{t}, \quad \mathbf{X}^{(3)}(t) = \begin{pmatrix} 2 \\ 1 \\ 1 \end{pmatrix} e^{2t},$$

and the general solution is

$$\mathbf{x}(t) = c_1 \begin{pmatrix} -1 \\ -2 \\ 1 \end{pmatrix} e^{-t} + c_2 \begin{pmatrix} 0 \\ 0 \\ 1 \end{pmatrix} e^{t} + c_3 \begin{pmatrix} 2 \\ 1 \\ 1 \end{pmatrix} e^{2t}.$$

Suppose we are seeking the specific solution with initial value

$$\mathbf{x}(0) = \begin{pmatrix} 3 \\ 5 \\ 0 \end{pmatrix}.$$

The constants c_1, c_2, and c_3 must satisfy

$$c_1 \begin{pmatrix} -1 \\ -2 \\ 1 \end{pmatrix} + c_2 \begin{pmatrix} 0 \\ 0 \\ 1 \end{pmatrix} + c_3 \begin{pmatrix} 2 \\ 1 \\ 1 \end{pmatrix} = \begin{pmatrix} -1 & 0 & 2 \\ -2 & 0 & 1 \\ 1 & 1 & 1 \end{pmatrix} \begin{pmatrix} c_1 \\ c_2 \\ c_3 \end{pmatrix} = \begin{pmatrix} 3 \\ 5 \\ 0 \end{pmatrix}.$$

To solve this linear system, we use the command **linsolve**. Let **m** be the coefficient matrix and let **b** be the right-hand side:

m := [[−1, 0, 2], [−2, 0, 1], [1, 1, 1]];   b := [3, 5, 0];

(Note that the columns of the matrix **m** are the eigenvectors of **A**.) Then to solve the system we type **linsolve(m, b)**, which produces $[-\frac{7}{3}, 2, \frac{1}{3}]$. Thus the solution of the linear system is $c_1 = -7/3$, $c_2 = 2$, $c_3 = 1/3$, and the solution to our initial value problem is

$$\mathbf{x}(t) = -\frac{7}{3} \begin{pmatrix} -1 \\ -2 \\ 1 \end{pmatrix} e^{-t} + 2 \begin{pmatrix} 0 \\ 0 \\ 1 \end{pmatrix} e^{t} + \frac{1}{3} \begin{pmatrix} 2 \\ 1 \\ 1 \end{pmatrix} e^{2t}.$$

EXAMPLE 2. Consider the system

$$\mathbf{x}' = \begin{pmatrix} 3 & -2 \\ 4 & -1 \end{pmatrix} \mathbf{x} \qquad \text{(Boyce \& DiPrima, Sect. 7.6, Prob. 1)}.$$

We set **B := [[3, −2], [4, −1]]**, and find the eigenpairs with **eigenvects(B)**, which produces

$$[1 + 2I, 1, \{[1, 1 - I]\}], [1 - 2I, 1, \{[1, 1 + I]\}]$$

So the eigenpairs are

$$1 + 2i, \begin{pmatrix} 1 \\ 1 - i \end{pmatrix}$$

and

$$1 - 2i, \quad \begin{pmatrix} 1 \\ 1 + i \end{pmatrix}.$$

Note that the second pair is the complex conjugate of the first. Thus a complex fundamental set is

$$\mathbf{X}^{(1)}(t) = \begin{pmatrix} 1 \\ 1 - i \end{pmatrix} e^{(1+2i)t}, \quad \mathbf{X}^{(2)}(t) = \begin{pmatrix} 1 \\ 1 + i \end{pmatrix} e^{(1-2i)t}.$$

The real and imaginary parts of the solution can be extracted by hand, or by using the *Maple* commands **evalm, evalc, Re, Im** and the composition operator **@** as follows:

```
csol := evalm(exp((1 + 2*I)*t)*[1, 1 − I]);
map(evalc@Re, csol);
map(evalc@Im, csol);
```

(See the glossary or the online help for more information on the above commands.) Either way, we see that a fundamental set of real solutions is

$$\mathbf{u}(t) = e^t \left[\begin{pmatrix} 1 \\ 1 \end{pmatrix} \cos 2t + \begin{pmatrix} 0 \\ 1 \end{pmatrix} \sin 2t \right]$$

$$\mathbf{v}(t) = e^t \left[- \begin{pmatrix} 0 \\ 1 \end{pmatrix} \cos 2t + \begin{pmatrix} 1 \\ 1 \end{pmatrix} \sin 2t \right].$$

EXAMPLE 3. Consider the system

$$\mathbf{x}' = \begin{pmatrix} 3 & -4 \\ 1 & -1 \end{pmatrix} \mathbf{x} \qquad \text{(Boyce \& DiPrima, Sect. 7.7, Prob. 1).}$$

We set **A := [[3, −4], [1, −1]]**. The command **eigenvects(A)** produces

$$[1, 2, \{[2, 1]\}]$$

Though the eigenvalue 1 has multiplicity two, it only has one linearly independent eigenvector. Thus $\rho = 1$ and $\boldsymbol{\xi} = \begin{pmatrix} 2 \\ 1 \end{pmatrix}$, and

$$\mathbf{X}^{(1)}(t) = \begin{pmatrix} 2 \\ 1 \end{pmatrix} e^t$$

is one solution. To find a second solution we solve

$$(\mathbf{A} - \rho\mathbf{I})\boldsymbol{\eta} = \begin{pmatrix} 2 & -4 \\ 1 & -2 \end{pmatrix} \boldsymbol{\eta} = \boldsymbol{\xi} = \begin{pmatrix} 2 \\ 1 \end{pmatrix}.$$

Let **m := [[2, −4], [1, −2]]** be the coefficient matrix, and let **b := [2, 1]** be the right-hand side. Then **linsolve(m, b)** produces

$$[1 + 2\_t_1, \_t_1]$$

We can choose one of the infinitely many solutions by taking $t_1 = 0$, so

$$\eta = \begin{pmatrix} 1 \\ 0 \end{pmatrix}.$$

Thus, a second solution is

$$\mathbf{X}^{(2)}(t) = \xi t e^{\rho t} + \eta e^{\rho t} = \begin{pmatrix} 2 \\ 1 \end{pmatrix} t e^t + \begin{pmatrix} 1 \\ 0 \end{pmatrix} e^t.$$

The general solution is

$$\mathbf{X}(t) = c_1 \begin{pmatrix} 2 \\ 1 \end{pmatrix} e^t + c_2 [\begin{pmatrix} 2 \\ 1 \end{pmatrix} t e^t + \begin{pmatrix} 1 \\ 0 \end{pmatrix} e^t].$$

We have shown how to use *Maple*'s linear algebra commands to find eigenpairs, and thus solutions, for linear systems of differential equations with constant co-efficients. The solutions to systems of linear equations can also be found with a direct application of **dsolve**.

EXAMPLE. Consider the system $x' = y$, $y' = -x$. To find its general solution, you would type

> **dsolve({diff(x(t), t) = y(t), diff(y(t), t) = −x(t)}, {x(t), y(t)});**

The result is

$$x(t) = c_1 \sin t + c_2 \cos t,$$

$$y(t) = c_1 \cos t - c_2 \sin t.$$

To solve the same system with initial conditions $x(0) = 1$, $y(0) = 0$, you would type

> **dsolve({diff(x(t), t) = y(t), diff(y(t), t) = −x(t), x(0) = 1, y(0) = 0}, {x(t), y(t)});**

Both methods for solving linear systems are useful. Solving in terms of the eigenpairs of the coefficient matrix is somewhat involved, but yields a simple and useful formula for the solution. The eigenpairs contain valuable information about the solution, which we will exploit further in Problem Set F. On the other hand, it is simpler to use **dsolve**, and for many purposes the solutions generated by **dsolve** are completely satisfactory.

Here are some examples on which you can practice:

(a) $x' = -3x + 2y$, $y' = -x$
(b) $x' = x - 2y$, $y' = -x$
(c) $x' = x + 2y$, $y' = -x$.

Phase Portraits

A solution of a 2×2 linear system is a pair of funtions $x(t)$, $y(t)$. The plot of a solution as a function of t would be a curve in (t, x, y)-space. Generally, such 3-dimensional plots are too complicated to be illuminating. Therefore, we project the curve from 3-dimensional space into the (x, y)-plane. This plane is called the *phase plane*, and the resulting curve in the phase plane is called a *trajectory*. A trajectory is drawn by plotting $(x(t), y(t))$ as the parameter t varies. A plot of a family of trajectories is called a *phase portrait* of the system.

A trajectory is an example of a *parametrized curve*. The *Maple* **plot** command will automatically display **[x(t), y(t), t = a..b]** as a parametric curve. In this section, we describe how to use *Maple*: (1) to solve an initial value problem consisting of a linear system and an initial condition, and plot the corresponding trajectory; (2) to solve a linear system using **dsolve** and draw a phase portrait; (3) to solve a system numerically and draw a phase portrait.

Plotting a Single Trajectory. Consider the linear system in Example (a) above, with initial condition $x(0) = 1$, $y(0) = 0$. We want to plot the trajectory with this initial condition. First we enter the initial value problem.

> **ivp := {diff(x(t), t) = −3*x(t) + 2*y(t), diff(y(t), t) = −x(t), x(0) = 1, y(0) = 0};**

Then we type the following:

> **solivp := dsolve(ivp, {x(t), y(t)});**
> **parasol := unapply(subs(solivp, [x(t), y(t), t = trange]), trange);**
> **traj := trange −> plot(parasol(trange));**

This program does two things. First, it produces the solution, **solivp**, of the initial value problem. This solution is a set of equations describing the dependent variables x and y as functions of t. The parametric solution, **parasol**, converts the form of the solution into parametric equations that can be fed into the **plot** command. Next, it defines a function **traj** that plots the trajectory corresponding to **solivp**. The function **traj** takes one argument, a range specifying the endpoints of the time interval for the trajectory. Figure 1 contains the picture generated by the command **traj(−0.3..5)**. To use this program to plot the solution of a different initial value problem, all you have to do is modify the definition of **ivp** appropriately.

Plotting Several Trajectories. Consider the linear system in Example (b) above. We want to plot trajectories of this system with various initial conditions $x(0) = a$, $y(0) = b$. We can do this by modifying the program above to solve this system with initial conditions $x(0) = a$, $y(0) = b$.

> **ivp := {diff(x(t), t) = x(t) − 2*y(t), diff(y(t), t) = −x(t), x(0) = a, y(0) = b};**
> **solivp := dsolve(ivp, {x(t), y(t)});**
> **parasol := unapply(subs(solivp, [x(t), y(t), t = trange]), a, b, trange);**

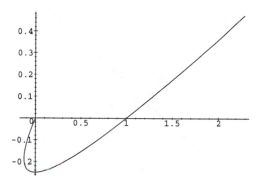

Figure 1

The function **parasol** depends on the initial values a and b and the range of the independent variable. For example, **parasol(1, 2, −3..3)** is a parametric curve, defined for $-3 \leq t \leq 3$ and determined by the initial values $x(0) = 1$, $y(0) = 2$. Next we define a function **phase** that plots a phase portrait of the linear system. We do this by creating a set of parametric solutions with different initial conditions and passing them to the **plot** command.

phase := trange –> plot({seq(seq(parasol(a, b, trange), a = −2..2), b = −2..2)});

Note that we have chosen a rectangular grid of 25 initial conditions, with a ranging from -2 to 2 and b ranging from -2 to 2 (in integer increments).

To use the function **phase**, simply enter **phase(t0..t1)**, where $t0$ and $t1$ are the endpoints of the time interval $[t0, t1]$ for which you wish to plot the trajectories. This will plot 25 trajectories corresponding to the 25 different initial conditions. The time interval $[t0, t1]$ must be chosen somewhat carefully; a small time interval could result in solution curves that are too short to get an idea of where they're headed, while a large time interval could cause the scale of the graph to get very large, resulting in solution curves that are squeezed together. Moreover, the larger the time interval, the longer it will take for *Maple* to produce the plot. You should start with a small time interval like $[-1, 1]$, and then adjust the interval to get a satisfactory picture. The command **phase(−2..1)** generated the phase portrait shown in Figure 2.

You should be able to plot the phase portraits for any 2×2 linear system with constant coefficients using the procedure above, replacing the differential equations in the definition of **ivp**, and choosing, by trial and error, an appropriate time interval to use with the function **phase**. By combining the phase portrait with a plot of the vector field (see Chapter 13), you should be able to determine the direction of increasing t on the curves. Also, in each case you should be able to decide whether the origin is a *sink* (all solutions approach the origin as t increases), *source* (all

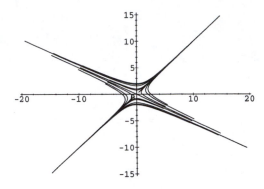

Figure 2

solutions move away from the origin), *saddle point* (solutions approach the origin along one direction, then move away from the origin in another direction), or *center* (solutions follow closed curves around the origin). The procedure for plotting a phase portrait must be modified slightly if you wish to vary only one parameter in the initial data. For example, the program

```
ivp := {diff(x(t), t) = x(t) + 2*y(t), diff(y(t), t) = −x(t), x(0) = a, y(0) = 0};
solivp := dsolve(ivp, {x(t), y(t)});
parasol := unapply(subs(solivp, [x(t), y(t), t = trange]), a, trange);
phase := trange −> plot({seq(parasol(a, trange), a = −4..4)});
```

will yield (upon entering **phase(−4..3)**) the nine curves shown in Figure 3 as a phase portrait for Example (c). (Count the curves! One of the solutions is the equilibrium solution $x(t) = 0$, $y(t) = 0$.)

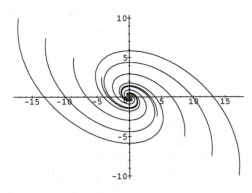

Figure 3

Plotting Trajectories of Numerical Solutions. The symbolic solver **dsolve** should be able to solve any homogeneous 2×2 linear system with constant coefficients. For inhomogeneous linear systems, linear systems with variable coefficients, and nonlinear systems, **dsolve** will generally not do the job. In these situations, we must turn either to the numerical solver **dsolve(..., numeric)** or to **DEplot**.

The **DEplot** command (which is in the **DEtools** package) was introduced in Chapter 7 for plotting numerical solutions to a single differential equation. (See also Chapter 8 for some additional hints about using **DEplot**.) With a slightly different syntax, **DEplot** can also be used for systems of differential equations. For example, you can type

> **with(DEtools):**
> **des := diff(x(t), t) = x(t) − 2*y(t), diff(y(t), t) = −x(t);**
> **iniset := {seq(seq([x(0) = a, y(0) = b], a = −2..2), b = −2..2)};**
> **pphase := trange −> DEplot([des], [x(t), y(t)], t = trange, iniset, stepsize = 0.1,**
> **method = rkf45, linecolor = black, arrows = NONE);**

Once **pphase** has been defined as above, the command **pphase(−2..1)** will produce a phase portrait similar to Figure 2. To see a direction field in addition to the solution curves, you can omit the option **arrows = NONE** or specify a different type of arrows if you prefer.

To limit the x and y ranges of the graph, say to $[-5, 5]$ in each direction, you can use the **DEplot** options **x = −5..5** and **y = −5..5**. When you do so, there is an important choice to be made—should **DEplot** continue to follow numerical trajectories after they leave the specified rectangle, in case they later return? The default behavior of **DEplot** is to stop computing a trajectory once it leaves the rectangle. This can make the command run much faster and, in the case of nonlinear equations, can sometimes avoid error messages due to trajectories that blow up in finite time. For example, the command

> **DEplot([des], [x(t), y(t)], t = −10..10, iniset, x = −5..5, y = −5..5, stepsize = 0.1,**
> **method = rkf45, linecolor = black, arrows = NONE);**

should produce a good phase portrait in a reasonable amount of time even though the t range specified, $[-10, 10]$, is much larger than it needs to be for the given x and y ranges. However, in some cases it is necessary to override the default behavior with the option **obsrange = FALSE** in order to get an accurate phase portrait—for instance, when the trajectories spiral around an equilibrium, or when some of the initial conditions lie outside the range of the graph.

Finally, we note that **DEplot** also works for systems of more than two equations, and for time-dependent (nonautonomous) systems. An important option in these cases is **scene**, which is used to specify a pair of variables to plot. For example, you can get a plot of x versus t for a trajectory of the above system by typing

> **DEplot([des], [x(t), y(t)], t = −3..3, {[x(0) = 1, y(0) = 1]}, scene = [t, x], x = −5..5,**
> **stepsize = 0.1, method = rkf45, linecolor = black, arrows = NONE);**

In the above example it is necessary to specify **arrows = NONE** in order to avoid an error message.

REMARK. *Maple* has a **phaseportrait** command (in the **DEtools** package) for plotting phase portraits; however, it is basically a restricted version of **DEplot** with less functionality. We recommend using **DEplot** rather than **phaseportrait**.

If the numerical values of a solution are needed, then **dsolve(. . ., numeric)** must be used. The following commands make a table of the x and y coordinates of the solution with initial conditions $x(0) = 1, y(0) = 1$ for $t = 0, 1, 2, \ldots, 10$.

```
numsol := dsolve({des, x(0) = 1, y(0) = 1}, {x(t), y(t)}, numeric,
    maxfun = 1000, startinit = TRUE);
array([seq(numsol(u), u = 0..10)]);
```

Though it is possible to produce phase portraits with **dsolve(. . ., numeric)** and **plot**, the procedure is cumbersome. It is simpler to use **DEplot**, and with the option **method = rkf45**, the same underlying numerical method is used.

Chapter 13

Qualitative Theory for Systems of Differential Equations

In this chapter, we extend the qualitative theory of autonomous differential equations from single equations to systems of two equations. Thus, consider a system of the form

$$\begin{cases} x' = F(x, y) \\ y' = G(x, y). \end{cases} \tag{1}$$

We assume the functions F and G have continuous partial derivatives everywhere in the plane, and the critical points—the common zeros of F and G—are isolated.

We know from the fundamental existence and uniqueness theorems that, corresponding to any choice of initial data (x_0, y_0), there is a unique pair of functions $x(t)$, $y(t)$ that satisfy the system (1) of differential equations and the initial conditions $x(0) = x_0$, $y(0) = y_0$. Taken together, this pair of functions parametrizes a curve, or trajectory, in the *phase plane* that passes through the point (x_0, y_0). Since we are dealing with autonomous systems (*i.e.*, the variable t does not appear on the right side of (1)), the solution curves are *independent of the starting time*. That is, if we consider initial conditions $x(t_0) = x_0$, $y(t_0) = y_0$, then we get the same solution curve with a time delay of t_0 units. Moreover, we also know that two distinct trajectories cannot intersect. Thus, the plane is covered by the family of trajectories; the plot of these curves is the *phase portrait* of (1).

As we know, we can find explicit formula solutions $x(t)$, $y(t)$ only for simple systems. Thus, we are forced to turn to qualitative and numerical methods. It is our purpose here to discuss qualitative techniques for studying the solutions of (1).

In most cases of physical interest, every solution curve behaves in one of the following ways:

(i) The curve consists of a single critical point;

(ii) The curve tends to a critical point as $t \to \infty$;

(iii) The curve is unbounded: the distance from $(x(t), y(t))$ to the origin becomes arbitrarily large as $t \to \infty$, or as $t \to t^*$, for some finite t^*;

(iv) The curve is periodic: the parametric functions satisfy $x(t + t_p) = x(t)$, $y(t + t_p) = y(t)$ for some fixed t_p;

(v) The curve approaches a periodic solution, *e.g.*, spiraling in on a circle.

Qualitative techniques may be able to identify the kinds of solutions that appear. In order to keep matters simple, we try to answer the following specific questions:

(1) What can we say about the critical points? Can we make a qualitative guess about their nature and stability? Can we deduce anything about the solution curves that start out close to the critical points?

(2) Can we predict anything about the long-term behavior of the solution curves? This includes those near critical points as well as the solution curves in general.

The list of possible limiting behaviors of solution curves is based on the Poincaré-Bendixson Theorem (*cf.* Theorem 9.7.3 in Boyce & DiPrima), which is valid for autonomous systems of two equations. Systems of three or more equations can have much more complicated limiting behavior. For example, solution curves can remain bounded and yet fail to approach any equilibrium or periodic state as $t \to \infty$. This phenomenon, called "chaos" by mathematicians, was anticipated by Poincaré (and perhaps even earlier by the physicist Maxwell), but most scientists did not appreciate how widespread chaos is until the arrival of computers. See Problem 13 in Problem Set F for an example.

There are two qualitative methods for analyzing systems of equations: one based on the idea of a *vector field*; and the other based on *linearized stability analysis*. The latter method is treated in detail in Sections 9.3–9.5 of Boyce & DiPrima; it provides information about the stability of critical points of a nonlinear system by studying associated linear systems. In this chapter, we focus on the former method: the use of vector fields in qualitative analysis. Some of the problems in Problem Set F involve linearized stability analysis.

Here is the basic idea. For any solution curve $(x(t), y(t))$, and any point t, we have the differential equations

$$\begin{cases} x'(t) = F(x(t), y(t)) \\ y'(t) = G(x(t), y(t)). \end{cases}$$

If we employ vector notation

$$\mathbf{x}(t) = \begin{pmatrix} x(t) \\ y(t) \end{pmatrix}, \quad \mathbf{f}(\mathbf{x}) = \begin{pmatrix} F(\mathbf{x}) \\ G(\mathbf{x}) \end{pmatrix},$$

then the system is written

$$\mathbf{x}' = \mathbf{f}(\mathbf{x}).$$

Moreover, the vector $\mathbf{x}'(t) = \mathbf{f}(\mathbf{x}(t))$ is just the tangent vector to the curve $\mathbf{x}(t)$. Since knowledge of the collection of tangent vectors to the solution curves would give a good idea of the curves themselves, it would be useful to have a plot of these vectors. Thus, corresponding to any vector $\mathbf{x}$, we draw the vector $\mathbf{f}(\mathbf{x})$, translated so that its foot is at the point $\mathbf{x}$. This plot is called the vector field of the system (1). Since we cannot do this at every point of the phase plane, we only draw the vectors at a set of regularly chosen points. Then we step back and look at the resulting vector field. We do not have the solution curves sketched, but we do have a representative collection of their tangent vectors. From the tangent vectors, we can get a good idea of the curves themselves. Indeed, we can often answer the questions above by a careful examination of the vector field.

Vector fields can be drawn by hand for simple systems, but the command **fieldplot** can draw vector fields of any first order system of two equations. We illustrate the use of this command by examining two specific systems, namely

$$\begin{cases} x' = x(1 - x - y) \\ y' = y(0.75 - 0.5x - y); \end{cases} \tag{2}$$

and

$$\begin{cases} x' = x(5 - x - y) \\ y' = y(-2 + x). \end{cases} \tag{3}$$

System (2) is a competing species model that is discussed in Boyce & DiPrima in Section 9.4, Example 1. System (3) is a predator-prey model from Section 7.4 of Edwards & Penney, **Elementary Differential Equations with Applications**, 3rd edition.

The **fieldplot** command is contained in the **plots** package. The syntax of the command is

fieldplot([F(x, y), G(x, y)], x = x0..x1, y = y0..y1);

We first consider system (2). To draw the vector field we enter

fieldplot([x*(1 − x − y), y*(0.75 − 0.5*x − y)], x = 0..2, y = 0..1);

The result is shown in Figure 1.

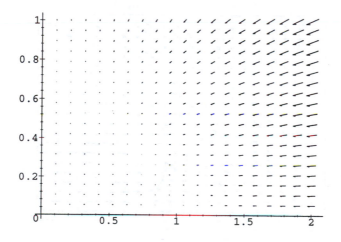

Figure 1

In a vector field, the vector drawn at the point $\mathbf{x} = (x, y)$ indicates both the direction and length of $\mathbf{f}(\mathbf{x})$. The vector at $\mathbf{x}$ drawn by **fieldplot** faithfully represents the direction of $\mathbf{f}(\mathbf{x})$ but, for the sake of visual effectiveness, its length is only proportional to the length of $\mathbf{f}(\mathbf{x})$. The critical, or equilibrium, points are those points at which the vector field vanishes, *i.e.*, those points where $F(x, y) = G(x, y) = 0$. Thus, the vectors are very short near the critical points. In the plot above we can approximately identify the critical points, but it is difficult to see the directions of the very short vectors near the critical points.

A simple way to see the directions of vectors near the critical points is to replace the **fieldplot** command with the **dfieldplot** command. This command is contained in the **DEtools** package; we used it previously to plot the direction field of a single first order differential equation. It produces plots in which every vector has the same length. The drawback, of course, is that it may be harder to locate the critical points (since there are no short arrows); the compensating benefit is that you can see the directions more clearly.

Here is the command we use to plot the "equal length" vector field.

> **dfieldplot([diff(x(t), t) = x(t)*(1 − x(t) − y(t)), diff(y(t), t) =**
> **y(t)*(0.75 − 0.5*x(t) − y(t))], [x(t), y(t)], t = 0..1, x = 0..1.5, y = 0..1,**
> **arrows = SLIM, axes = BOXED);**

(The third argument **t = 0..1** is irrelevant to the vector field plotted in this example, but the syntax of **dfieldplot** requires that some range be specified for the independent variable **t**.) The result is shown in Figure 2.

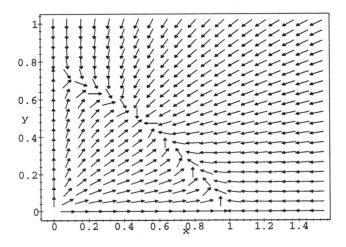

Figure 2

This picture gives us a much clearer indication of what's going on, but it still doesn't allow us to identify the critical points precisely. To do this we actually have to solve the equations $x' = 0$ and $y' = 0$ simultaneously. We can do this with the **solve** command,

solve({x*(1 − x − y) = 0, y*(0.75 − 0.5*x − y) = 0}, {x, y});

which produces the answer

$$\{x = 0, y = 0\}, \{x = 1., y = 0\}, \{x = 0, y = .7500000000\},$$
$$\{y = .5000000000, x = .5000000000\}$$

Thus, the critical points are $(0,0)$, $(0, 0.75)$, $(1, 0)$, and $(0.5, 0.5)$. Knowing the critical points, we can deduce from the vector field that $(0,0)$ is an unstable node, $(0, 0.75)$ and $(1, 0)$ are unstable saddle points, and $(0.5, 0.5)$ is an asymptotically stable node. Solutions starting near the saddle points (but not on the axes) tend away from them, and those starting near the point $(0.5, 0.5)$ tend toward it. In fact, the vector field strongly suggests that every solution curve starting in the first quadrant (but not on the axes) tends toward $(0.5, 0.5)$. Hence $(0.5, 0.5)$, which corresponds to equal populations, is apparently the limiting state that all positive solutions approach as t increases. We have thus answered both questions (1) and (2) on the basis of the vector field. Figure 3 provides a closer look at the vector field near the point $(0.5, 0.5)$.

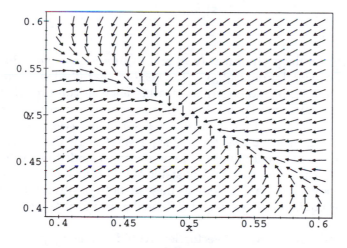

Figure 3

Now consider the predator-prey system (3). Here $x(t)$ represents the population of prey and $y(t)$ the population of predators. The critical points of this system are

$(0,0), (5,0), (2,3)$. As in system (2), our goal is to use the vector field to answer questions (1) and (2). Here is the appropriate *Maple* command :

dfieldplot([diff(x(t), t) = x(t)*(5 − x(t) − y(t)), diff(y(t), t) = y(t)*(−2 + x(t))],
[x(t), y(t)], t = 0..1, x = 0..6, y = 0..4, arrows = SLIM, axes = BOXED);

The output is shown in Figure 4.

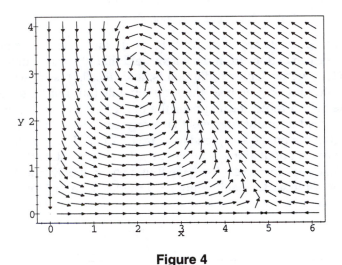

Figure 4

It is evident from the portrait that $(0,0)$ and $(5,0)$ are unstable saddle points and that solutions spiral around $(2,3)$. The vectors near $(2,3)$ suggest that the solutions spiral into $(2,3)$, and linearized stability analysis confirms that $(2,3)$ is an asymptotically stable spiral point. In particular, solutions starting near the first two critical points tend away from them, whereas solution curves starting near the latter tend toward it. In fact, the vector field suggests strongly that every solution curve starting in the first quadrant (but not on the axes) tends toward the critical point $(2,3)$. This is the strongest kind of stable equilibrium—namely, no matter what the initial populations are, as long as they are both positive, the population tends toward the equilibrium position of 40% predators and 60% prey.

Remarks. We indicated above that vector fields could be drawn by hand. It would be difficult, however, to draw as satisfactory a vector field as we have produced with *Maple*. On the other hand, we note that a careful study of the signs of F and G can lead to a good understanding of the nature of the vector field. This is done for the system of Example (2) in Section 9.4 in Boyce & DiPrima. See in particular Figure 9.4.5 there.

The essential idea in Chapter 6 for a single autonomous first order equation

$$x' = f(x)$$

is to determine the qualitative nature of the solutions purely on the basis of the sign of $f(x)$ for various x's. Actually, we have done the same thing here: namely, determine the qualitative nature of the solutions (trajectories) of a system from information on the signs of F and G.

As mentioned above, according to linearized stability analysis, the nature of critical points for nonlinear systems can be deduced from the associated linear systems. If the point is asymptotically stable for the associated linear system (the case if the eigenvalues are negative, or have negative real parts), then the point is asymptotically stable for the original nonlinear system. If it is unstable for the linear system, it is unstable for the nonlinear system. Centers are ambiguous. This analysis is valid in exactly the same way for a single equation. Suppose $\bar{x}$ is a critical point for $x' = f(x)$. Then letting $u(t) = x(t) - \bar{x}$, we see that

$$u' = f(u + \bar{x}) \approx f(\bar{x}) + f'(\bar{x})u = f'(\bar{x})u,$$

so the associated linear equation is $u' = f'(\bar{x})u$. The solutions are $u(t) = ce^{\lambda t}$, where $\lambda = f'(\bar{x})$. Therefore, $x(t) = \bar{x} + ce^{\lambda t}$. It is clear that $\bar{x}$ is asymptotically stable if $\lambda < 0$ and unstable if $\lambda > 0$. The same is true of the nonlinear system (with the situation for $\lambda = 0$ being ambiguous). Thus, we see that this assessment of stability parallels the situation for systems.

Conclusion. Solutions of nonlinear systems can be calculated and their trajectories plotted with **DEplot**. Suggestions on convenient ways to do this are made in Chapter 12. Taken together, the information provided by a vector field, by a plot of the trajectories, and by linearized stability analysis, gives a complete understanding of the behavior of the solutions of a system of two first order equations. The information provided by these three approaches is similar, but each approach yields information the other approaches don't provide. For example, a vector field provides a good global indication of what the phase portrait will look like, as well as indicating the approximate location of the critical points and their type; a plot of the trajectories gives a precise indication of the solution curves; and a linearized stability analysis determines the type of the critical points.

In using these approaches to study a system, it is almost always best to start with a vector field plot or with linearized stability analysis. Typically, it takes a long time to plot trajectories of nonlinear systems. But by using the information from a vector field plot or from linearized stability analysis you can choose appropriate initial data points to use with **DEplot**. A vector field plot will also help you choose an appropriate time interval (*e.g.*, positive or negative time).

Problem Set F

Systems of Equations

A number of suggestions for using *Maple* to solve these problems may be found in Chapter 12. The parameters in those suggestions may need to be modified according to the problem at hand. In particular, you may have to change the time interval or the set of initial conditions.

1. In this problem, we study three systems of equations taken from Boyce & DiPrima:

$$\mathbf{x}' = \begin{pmatrix} 2 & -1 \\ 3 & -2 \end{pmatrix} \mathbf{x} \qquad \text{(Prob. 3, Sect. 7.5)} \qquad (i)$$

$$\mathbf{x}' = \begin{pmatrix} 1 & -1 \\ 5 & -3 \end{pmatrix} \mathbf{x} \qquad \text{(Prob. 5, Sect. 7.6)} \qquad (ii)$$

$$\mathbf{x}' = \begin{pmatrix} -3 & 5/2 \\ -5/2 & 2 \end{pmatrix} \mathbf{x} \qquad \text{(Prob. 4, Sect. 7.7)}. \qquad (iii)$$

(a) Use *Maple* to find the eigenvalues and eigenvectors of each linear system. Use the eigenvectors and eigenvalues to write down the general solution of the system.

(b) Use **dsolve** to compute the solution of system (i). Compare the general solution you obtained in part (a) with the solution generated by **dsolve**. Are they the same? If not, explain how they are related (that is, match up the constants).

(c) Using the solution formulas obtained above, plot several trajectories of each system. On your graphs, identify the eigenvectors (if relevant), the direction of increasing time on the trajectories, and the type and stability of the origin as a critical point. You may find that the quality of your portraits can be enhanced by modifying the t-interval or the range of values assumed by the initial data. (A word of caution: If you request too many curves, *Maple* will take a long time to generate the plot.)

2. Do Problem 1 but replace the three systems listed there by the following three from Boyce & DiPrima:

$$\mathbf{x}' = \begin{pmatrix} 5/4 & 3/4 \\ 3/4 & 5/4 \end{pmatrix} \mathbf{x} \qquad \text{(Prob. 6, Sect. 7.5)} \qquad (i)$$

$$\mathbf{x}' = \begin{pmatrix} 2 & -5/2 \\ 9/5 & -1 \end{pmatrix} \mathbf{x} \qquad \text{(Prob. 4, Sect. 7.6)} \qquad (ii)$$

$$\mathbf{x}' = \begin{pmatrix} -3/2 & 1 \\ -1/4 & -1/2 \end{pmatrix} \mathbf{x} \qquad \text{(Prob. 3, Sect. 7.7)}. \qquad (iii)$$

3. Use the command **eigenvects** to find the eigenvalues and eigenvectors of the following systems. Use the eigenvalues and eigenvectors to write down the general solutions.

(a)

$$\mathbf{x}' = \begin{pmatrix} 2 & -1 \\ 1 & -2 \end{pmatrix} \mathbf{x}.$$

Let $\mathbf{x}(t) = (x(t), y(t))$. Determine the possible limiting behavior of $x(t)$, of $y(t)$, and of $x(t)/y(t)$ as t approaches $+\infty$.

(b)

$$\mathbf{x}' = \begin{pmatrix} 4 & 3 \\ -3 & -2 \end{pmatrix} \mathbf{x}.$$

(c)

$$\mathbf{x}' = \begin{pmatrix} 0 & 0 & -1 \\ 2 & 0 & 0 \\ -1 & 2 & 4 \end{pmatrix} \mathbf{x}.$$

Find the solution with initial condition

$$\mathbf{x}(0) = \begin{pmatrix} 7 \\ 5 \\ 5 \end{pmatrix}.$$

(d) Now solve the initial value problem in (c) with **dsolve** and compare the answer with the solution you obtained in (c).

4. *Maple* can solve some inhomogeneous linear systems. Here are three such systems (taken from Boyce & DiPrima, Sect. 7.9, Problems 1, 3, & 10).

$$\mathbf{x}' = \begin{pmatrix} 2 & -1 \\ 3 & -2 \end{pmatrix} \mathbf{x} + \begin{pmatrix} e^t \\ t \end{pmatrix} \qquad (i)$$

$$\mathbf{x}' = \begin{pmatrix} 2 & -5 \\ 1 & -2 \end{pmatrix} \mathbf{x} + \begin{pmatrix} -\cos t \\ \sin t \end{pmatrix} \qquad (ii)$$

$$\mathbf{x}' = \begin{pmatrix} -3 & \sqrt{2} \\ \sqrt{2} & -2 \end{pmatrix} \mathbf{x} + \begin{pmatrix} e^{-t} \\ -e^{-t} \end{pmatrix}. \tag{iii}$$

(a) Solve the three systems using **dsolve**.

(b) Now impose the initial data

$$\mathbf{x}(0) = \begin{pmatrix} 1 \\ 1 \end{pmatrix}$$

on each system. For each of (*i*)–(*iii*), draw the solution curve using a time interval $-1 \le t \le 1$. Based on the differential equation, indicate the direction of motion through the initial value. Then expand the time interval and, purely on the graphical evidence, venture a guess as to the behavior of the solution curve as $t \to -\infty$ and as $t \to \infty$.

5. Here we reconsider the pendulum models examined in Problems 3–6 of Problem Set D. (See also the discussion in Section 9.3 of Boyce & DiPrima.)

(a) Consider first the undamped pendulum

$$\theta'' + \sin \theta = 0, \quad \theta(0) = 0, \quad \theta'(0) = b.$$

Let $x = \theta$ and $y = \theta'$; then x and y satisfy the system

$$\begin{cases} x' = y \\ y' = -\sin x \end{cases} \qquad \begin{cases} x(0) = 0 \\ y(0) = b. \end{cases}$$

Solve this system numerically and plot, on a single graph, the resulting trajectories for the initial conditions $x(0) = 0$, and $y(0) = 0.5, 1, 1.5, 2, 2.5$. Use a positive time range (*e.g.*, **t = 0..20**) and the option **scaling = CONSTRAINED** to improve your pictures.

(b) Based on your pictures in part (a), describe physically what the pendulum seems to be doing in the three cases $\theta'(0) = 1.5, 2$, and 2.5.

(c) The energy of the pendulum is defined as the sum of kinetic energy $(\theta')^2/2$ and potential energy $1 - \cos \theta$, so

$$E = \frac{1}{2}(\theta')^2 + 1 - \cos \theta = \frac{1}{2}y^2 + 1 - \cos x.$$

Show, by taking the derivative dE/dt, that E is constant when $\theta(t)$ is a solution to the pendulum equation. What basic physical principle does this represent?

(d) Use **contourplot** (with the option **scaling = CONSTRAINED**, as in part (a)) to plot the level curves for the energy. Explain how your picture is related to the trajectory plot from part (a).

(e) If the pendulum reaches the upright position (*i.e.*, $\theta = \pi$), what must be true of the energy? Now explain why there is a critical value E_0 of the energy, below which the pendulum swings back and forth without reaching the upright position, and above which it swings overhead and continues to revolve in the same direction. What is E_0? What must the pendulum do when $E = E_0$? Explain. Next, consider the solution curve corresponding to initial data $\theta(0) = 0, \theta'(0) = b$. What is the value of the energy on that curve? Do your conclusions from part (b) agree with your analysis in this part?

(f) Now consider the damped pendulum

$$\theta'' + 0.5\theta' + \sin\theta = 0, \quad \theta(0) = 0, \quad \theta'(0) = b.$$

Numerically solve the corresponding first order system (which you must determine) and plot the resulting trajectories for the initial velocities $b = 0, 0.5, 1, \ldots, 6$. You will notice that all the trajectories in your graph tend toward the critical points $(0,0)$ or $(2\pi, 0)$. Explain what this means physically. There is a value b_0 for which the trajectory tends toward the critical point $(\pi, 0)$. Estimate, up to two-decimal accuracy, the value of b_0. What would this correspond to physically?

(g) Recall the energy defined in part (c). Compute E' and determine which of the following are possible: (1) E is increasing, (2) E is decreasing, (3) E stays the same. Explain how the possibilities are reflected in the solutions from part (f).

6. Consider the *competing species* model (Boyce & DiPrima, Prob. 2, Sect. 9.4)

$$\begin{cases} \dfrac{dx}{dt} = x(1.5 - x - 0.5y) \\ \dfrac{dy}{dt} = y(2 - 1.5x - 0.5y) \end{cases}$$

(a) Find all critical points of the system. At each critical point, calculate the corresponding linear system and find the eigenvalues of the coefficient matrix; then identify the type and stability of the critical point.

(b) Plot the vector field.

(c) Use several initial data points (x_0, y_0) in the first quadrant to draw a phase portrait for the system. Identify the direction of increasing t on the trajectories you obtain. Use your vector field plot from part (b) to choose a representative sample of initial conditions. Then combine the vector field and phase portrait on a single graph.

(d) Suppose the initial state of the population is given by

$$x(0) = 2.5, \ y(0) = 2.$$

Find the state of the population at $t = 1, 2, 3, 4, 5, \ldots, 20$.

(e) Explain why practically speaking there is no "peaceful coexistence"; *i.e.*, with the exception of an atypical set of starting populations (the *separatrix* curves), one or the other population must die out. For which nonzero initial populations is there no change? Sketch (by hand) on your plot the separatrices; *i.e.*, the solution curves that approach the unstable equilibrium point where both populations are positive and form the boundary between the solution curves that approach each of the two stable points where only one population survives. (In Problems 11 and 12, we investigate how to approximate separatrices using *Maple*.)

(f) The vertical line $x = 1.5$ cuts the separatrix. By experimenting with the long-term behavior of solutions emanating from several points on this line, find an approximation to two-decimal point accuracy of the value $\bar{y}$ such that $(1.5, \bar{y})$ is on the separatrix. Could you verify numerically whether the point is in fact on the separatrix? Why or why not?

7. Consider the *competing species* model (Boyce & DiPrima, Prob. 4, Sect. 9.4)

$$\begin{cases} \dfrac{dx}{dt} = x(1.5 - 0.5x - y) \\ \dfrac{dy}{dt} = y(0.75 - 0.125x - y) \end{cases}$$

(a) Find all critical points of the system. At each critical point, calculate the corresponding linear system and find the eigenvalues of the coefficient matrix; then identify the type and stability of the critical point.

(b) Plot the vector field.

(c) Use several initial data points (x_0, y_0) in the first quadrant to draw a phase portrait for the system. Identify the direction of increasing t on the trajectories you obtain. Use your vector field plot from part (b) to choose a representative sample of initial conditions. Then combine the vector field and phase portrait on a single graph.

(d) Suppose the initial state of the population is given by

$$x(0) = 0.1, \; y(0) = 0.1.$$

Find the state of the population at $t = 1, 2, 3, 4, 5, \ldots, 20$.

(e) Explain why, practically speaking, "peaceful coexistence" is the only outcome; *i.e.*, with the exception of the situation in which one or both species starts out without any population, the population distributions always tend toward a certain equilibrium point. Sketch (by hand) on your plot the separatrices that connect the stable equilibrium point to the two unstable points at which one population is zero; these separatrices divide the solution curves that tend toward the origin as $t \to -\infty$ from those that are unbounded as $t \to -\infty$. (In Problems 11 and 12, we investigate how to approximate separatrices using *Maple*.)

(f) The vertical line $x = 2.5$ cuts a separatrix. By experimenting with the behavior of solutions for *negative* time emanating from several points on this line, find an approximation to two-decimal point accuracy of the value $\bar{y}$ such that $(2.5, \bar{y})$ is on the separatrix. (*Hint*: $-3 \le t \le 0$ should suffice.) Could you verify numerically whether the point is in fact on the separatrix? Why or why not?

8. Consider the *predator-prey* model

$$\begin{cases} \dfrac{dx}{dt} = x(2-y) \\ \dfrac{dy}{dt} = y(x-1) \end{cases}$$

in which x represents the population of the prey and y represents the population of the predators.

(a) Find all critical points of the system. At each critical point, calculate the corresponding linear system and find the eigenvalues of the coefficient matrix; then identify the type and stability of the critical point.

(b) Plot the vector field.

(c) Use several initial data points (x_0, y_0) in the first quadrant to draw a phase portrait for the system. (Use the option **obsrange = FALSE** with **DEplot** because solutions may spiral out of and then back into the range of your graph.) Identify the direction of increasing t on the trajectories you obtain. Use your vector field plot from part (b) to choose a representative sample of initial conditions. Then combine the vector field and phase portrait on a single graph.

(d) Explain from your phase portrait how the populations vary for initial data close to the equilibrium point $(\bar{x}_0, \bar{y}_0)$ (the unique critical point inside the first quadrant). What about initial data far from the critical point?

(e) Suppose the initial state of the population is given by

$$x(0) = 1, \; y(0) = 1.$$

Find the state of the population at $t = 1, 2, 3, 4, 5$.

(f) Estimate the period of the solution curve with initial data $(1,1)$.

9. Consider a *predator-prey* model where the behavior of the prey is governed by a logistic equation (in the absence of the predator). Such a model is typified by the following system (Boyce & DiPrima, Prob. 3, Sect. 9.5):

$$\begin{cases} \dfrac{dx}{dt} = x(1 - 0.5x - 0.5y) \\ \dfrac{dy}{dt} = y(-0.25 + 0.5x) \end{cases}$$

where x represents the population of the prey, and y represents the population of the predators.

(a) Find all critical points of the system. At each critical point, calculate the corresponding linear system and find the eigenvalues of the coefficient matrix; then identify the type and stability of the critical point.

(b) Plot the vector field.

(c) Use several initial data points (x_0, y_0) in the first quadrant to draw a phase portrait for the system. (Use the option **obsrange = FALSE** with **DEplot** because solutions may spiral out of and then back into the range of your graph.) Identify the direction of increasing t on the trajectories you obtain. Use your vector field plot from part (b) to choose a representative sample of initial conditions. Then combine the vector field and phase portrait on a single graph.

(d) Explain from your phase portrait how the populations vary over time for initial data near the equilibrium point $(\bar{x}_0, \bar{y}_0)$ (the unique critical point inside the first quadrant, not on the axes). What happens when the initial condition is far from the equilibrium point?

(e) Suppose the initial state of the population is given by

$$x(0) = 1, \ y(0) = 1.$$

Find the state of the population at $t = 1, 2, 3, 4, 5$.

(f) Estimate how long it takes for both populations to arrive simultaneously within 0.01 of their equilibrium values if we start with initial data $(1, 1)$.

10. Consider a modified predator-prey system where the behavior of the prey is governed by a logistic/threshold equation (in the absence of the predator). Such a model is typified by the following system (Boyce & DiPrima, Prob. 5, Sect. 9.5):

$$\begin{cases} \dfrac{dx}{dt} = x(-1 + 2.5x - 0.3y - x^2) \\ \dfrac{dy}{dt} = y(-1.5 + x) \end{cases}$$

where x represents the population of the prey, and y represents the population of the predators.

(a) Find all critical points of the system. At each critical point, calculate the corresponding linear system and find the eigenvalues and eigenvectors of the coefficient matrix; then identify the critical points as to type and stability.

(b) Plot the vector field.

(c) Use several initial data points (x_0, y_0) in the first quadrant to draw a phase portrait for the system. (Use the option **obsrange = FALSE** with **DEplot** because solutions may spiral out of and then back into the range of your graph.) Use your vector field plot from part (b) to choose a representative sample of initial conditions.

(d) In parts (a), (b), and (c), you obtained information about the solutions of the system using three different approaches. Combine all of this information by combining the plots from parts (b) and (c) with the **display** command, and then drawing in by hand the critical points, the eigenvectors from part (a), and any separatrices. What information is provided by the approach in part (a) that is not provided by the other approaches? Answer the same question for parts (b) and (c).

(e) Interpret your conclusions in terms of the populations of the two species.

11. Some of the nonlinear systems of differential equations we have studied have more than one asymptotically stable equilibrium point. In such cases, it can be difficult to predict which (if any) equilibrium point the solution with a given initial condition will approach as time increases. Generally, the solution curves that approach one stable equilibrium will be separated from the solutions that approach another stable equilibrium by a separatrix curve. The separatrix is itself a solution curve that does not approach either stable equilibrium—often it approaches a saddle point instead. (See, for example, Figure 9.4.4 in Section 9.4 of Boyce & DiPrima.)

Having located a relevant saddle point, one can approximate the separatrix by choosing an initial condition very close to the saddle point and solving the differential equation *backwards* in time. The reason for going "back in time" is to find (approximately) a solution curve that approaches very close to the saddle point as time increases. For two population dynamics models, we will use this idea to get a fairly precise picture of which initial conditions go to which equilibria.

(a) Consider the system

$$\begin{cases} \dfrac{dx}{dt} = x(1.5 - x - 0.5y) \\ \dfrac{dy}{dt} = y(2 - 1.5x - 0.5y) \end{cases}$$

studied in Problem 6 above. There are two asymptotically stable equilibria, $(1.5, 0)$ and $(0, 4)$, a saddle point at $(1, 1)$, and an unstable node at $(0, 0)$. Plot a family of solution curves in the first quadrant (positive x and y) on the same graph; use a time interval from $t = 0$ to a positive time large enough that you can clearly see where the solutions curves are headed, but not so large that your plot takes forever to compute. (We suggest a set of initial values between 0.5 and 4.5. It also may be useful to adjust the range of the plot until you get a reasonably square graph.) Observe where the solution curves seem to be heading and where the separatrix seems to be.

(b) Draw an approximate separatrix by plotting a solution curve with initial values for x and y very close to the saddle point $(1,1)$, using a *negative* range of values for t. This should give you a good picture of the separatrix on one side of the saddle point; to approximate the rest of the separatrix you will need to choose a set of initial conditions on the other side of (but still very close to) the saddle point. In each case, you may need to fine-tune the time interval, since a time interval that is too small will only show you a small part of the separatrix, while too large an interval may cause the solution to blow up. Finally, show all of your plots on one graph, and make sure the separatrix really does separate the different asymptotic properties of the solution curves found in part (a). (To help in distinguishing the separatrix, you can use the option **linecolor = gray** in making your separatrix plots.) Discuss how the possible limiting values of the species depend on the initial values of the species.

12. Read the introduction to Problem 11.

(a) Do part (a) of Problem 11, but using the system

$$\begin{cases} \dfrac{dx}{dt} = x(-1 + 2.5x - 0.3y - x^2) \\ \dfrac{dy}{dt} = y(-1.5 + x) \end{cases}$$

from Problem 10 above. This time there are asymptotically stable equilibria at the origin and $(3/2, 5/3)$, and saddle points at $(0.5, 0)$ and $(2, 0)$.

(b) Do part (b) of Problem 11 for the system in part (a) of this problem. It is up to you to decide whether one or both of the saddle points are relevant to drawing the separatrix or separatrices that divide those solution curves that approach one equilibrium from those that approach the other as time increases; your picture from part (a) may be helpful. Also, keep in mind that since the saddle points are on the x-axis in this case, you need only consider nearby initial conditions above the x-axis because we are only interested in solutions in the first quadrant.

13. This problem is based on Problems 7.3.34–35 of Edwards and Penney, **Elementary Differential Equations with Applications**, 3rd edition.

(a) Consider the linear system

$$\begin{cases} \dfrac{dx}{dt} = -x + hy \\ \dfrac{dy}{dt} = x - y. \end{cases}$$

Draw phase portraits for the system for the values $h = 0, 0.5$, and -0.5. What can you deduce about the change in the type and/or stability of the equilibrium point $(0, 0)$ corresponding to relatively small perturbations of the system?

(b) Now consider the system

$$\begin{cases} \dfrac{dx}{dt} = y + hx(x^2 + y^2) \\ \dfrac{dy}{dt} = -x + hy(x^2 + y^2). \end{cases}$$

Draw phase portraits (using numerical solutions this time) for $h = 0, 1$, and -1. (We recommend that you do not use the option **method = rkf45** in this part because its use necessitates a very precise choice of the time interval for some of the phase portraits.) Illustrate how linearized stability analysis is inconclusive in the case of a *center* by citing the evidence in your portraits. (To decide which way the solution curves are going, you can examine the signs of dx/dt and dy/dt at a particular point, say $x = 1$, $y = 0$, or use *Maple* to plot the vector field in each case.)

14. In Chapter 5 we investigated the sensitivity to initial values of solutions of a single differential equation. In this problem, we investigate the same issue for two systems of equations. The systems we study are said to be *chaotic* because the solutions are sensitive to initial values, but in contrast to the examples in Chapter 5 and Problems 14 and 15 of Problem Set C, the solutions remain bounded.

Since we do not know the exact solutions for the systems we will study, in order to judge the time it takes for a small perturbation of the system to become large, we will compare pairs of numerical solutions whose initial conditions are close, but still relatively far apart compared with the local error (*cf.* Chapter 7) of our numerical method. To be safe we only consider perturbations as small as 10^{-4}.

(a) Consider the Lorenz system

$$x' = 10(y - x)$$
$$y' = 28x - y - xz$$
$$z' = -(8/3)z + xy$$

(which is studied, for example, in Section 9.8 of Boyce & DiPrima). We investigate the idea that small changes in the initial conditions can lead to large changes in the solution after a relatively short period of time. For $a = 0.1$, plot on the same graph the first coordinate $x(t)$ of the solutions corresponding to the initial conditions $(5, 5, 5)$ and $(5 + a, 5, 5)$ from $t = 0$ to $t = 20$. (We recommend that you use the **DEplot** command with **scene = [t, x]**. Since you will be using the initial condition $(5, 5, 5)$ again below, it may save time to use a separate **DEplot** command for each initial condition and combine the plots later with the **display** command. This approach also allows you to use an option like **linecolor = gray** in one of the **DEplot** commands in order to better distinguish between the two curves.) Observe the time at which the solutions start to differ noticeably. Repeat for $a = 0.01, 0.001, 0.0001$. (The criterion for a "noticeable" difference between the two solutions is up to you; just try to be consistent from one observation to the next.)

(b) Make a table or graph of the number of decimal places in which the initial condition was perturbed (that is, 1, 2, 3, and 4 for the four given values of a) versus the amount of time the solutions stayed close to each other. This graph shows roughly how long we should trust a numerical solution to be close to the actual solution for a given number of digits of accuracy in each step of our numerical method. Describe how the amount of time we can trust a numerical solution seems to depend on the number of digits of accuracy per time step, judging from your data. If a numerical method has 16 digits of accuracy, about how long do you think the numerical solution can be trusted? What if the accuracy were 100 digits?

15. Read the introduction to Problem 14.

   (a) Do part (a) of Problem 14, but using the Rössler system

$$x' = -y - z$$
$$y' = x + 0.36y$$
$$z' = 0.4x - 4.5z + xz.$$

   This time plot from $t = 0$ to $t = 100$ and use $(2, 2, 2)$ as the initial condition in place of $(5, 5, 5)$.

   (b) Do part (b) of Problem 14 for the Rössler system.

Complex Systems. In the following three problems, we consider a special class of 2×2 systems, namely, those that may be represented as a differential equation involving a single complex variable z. Consider, for example, the equation

$$z' = z(e^{i\phi} + z\bar{z}). \tag{1}$$

Here $z(t) = x(t) + iy(t)$ is a complex-valued function of the independent variable t, ϕ is a real parameter, and $\bar{z}$ is the complex conjugate of z. If we compute the real and imaginary parts of (1), we obtain the equivalent nonlinear 2×2 system

$$\begin{cases} x' = x \cos \phi - y \sin \phi + x^3 - xy^2 \\ y' = x \sin \phi + y \cos \phi + x^2 y - y^3. \end{cases} \tag{2}$$

   In spite of the fact that these are equivalent, the expression (1) offers several advantages over (2). First, (1) is "one-dimensional" in the sense that there is only one dependent variable z. Second, there are certain properties of the equation (1) that are apparent in the complex form, but obscure in the 2×2 form. For example, (1) is invariant by rotations through any angle (see the following problem), and therefore its set of solutions is rotationally symmetric. Symmetry plays a large role in the theory of differential equations, and complex-valued equations like (1) are the simplest kind of differential equations displaying rotational symmetry. Third, equations like (1) exhibit a phenomenon called *bifurcation* when the parameter ϕ varies. A parametrized family of differential equations is said to *bifurcate* at a parameter value ϕ_0 if the qualitative behavior of the differential equation changes as the parameter passes through ϕ_0. Examples of bifurcations are: loss of stability of a critical point; and creation or annihilation of limit cycles, or periodic solutions.

   You may have noticed that for certain real-valued differential equations **dsolve** returns complex-valued solutions. In fact, many *Maple* commands do not distinguish between real and complex numbers. This is true of the commands **dsolve** and **dsolve(. . ., numeric)**. In particular, we can use these commands on equations like (1) without worrying about the fact that these equations are complex-valued. *Maple* will report its solutions in exactly the same form as for real-valued differential equations, except that now the solutions will almost always be complex-valued (whereas in the case of real-valued equations, complex-valued solutions were rarely reported).

Unfortunately, the *Maple* command **DEplot** *does* distinguish between complex- and real-valued functions, so if we want to use **DEplot** to draw the phase portraits of the solutions in the complex plane, we first have to extract the real and imaginary parts of the complex equations, and then use **DEplot** on the real and imaginary parts. In order to save you a considerable amount of trouble figuring out exactly how to do this, we give you the following program:

```
with(DEtools):
complexsystem := proc(phi, trange) local x, y, z, zrhs, eqns, j, r, inits;
   zrhs := subs(z = x + I*y, z*(exp(I*phi) + z*conjugate(z)));
   eqns := [diff(x(t), t) = evalc(Re(zrhs)), diff(y(t), t) = evalc(Im(zrhs))];
   inits := {seq(seq([x(0) = r*cos(2*Pi*j/6), y(0) = r*sin(2*Pi*j/6)],
      r = [0.25, 0.75]), j = 0..5)};
   DEplot(eqns, [x(t), y(t)], t = trange, inits, x = -2..2, y = -2..2,
      stepsize = 0.1, arrows = NONE, linecolor = black);
end;
```

This program is designed to solve equation (1) numerically and plot solutions for a given value of the parameter ϕ. The other argument of the function specifies the time interval on which to solve the differential equation. For example, typing **complexsystem(Pi, −5..5)** will plot a phase portrait for $\phi = \pi$ using values of t from −5 to 5.

In order to use and modify this program, you should note the following features:

- The right-hand side of the complex differential equation (1) appears in the second line of the program, defining **zrhs**. Note that **I** is *Maple*'s notation for $i = \sqrt{-1}$.
- We have included initial values corresponding to the complex initial condition $z(0) = re^{i\theta}$, given in polar coordinates. Note that multiplication by $e^{i\theta}$ corresponds in the complex plane to counterclockwise rotation by the angle θ. The idea is to use a set of initial points which, like the differential equation, are rotationally symmetric around the origin (see parts (a) of the problems below). In this particular example, the initial conditions consist of points on the circles of radius 0.25 and 0.75, spaced evenly at angles $2\pi j/6$, for $j = 0, \ldots, 5$.
- We do not use the **DEplot** option **method = rkf45** because it makes **DEplot** more likely to produce an error message (and no graph) when solution curves approach infinity rapidly.

To summarize, the only things you'll have to modify to use this program in the following problems are:

- the actual differential equation used to define **zrhs**,
- the list of initial conditions **inits**, and
- possibly the x and y ranges in the **DEplot** command.

16. This exercise concerns the equation mentioned above,

$$z' = z(e^{i\phi} + z\bar{z}). \tag{i}$$

(a) Show that equation (i) is rotationally invariant in the following sense: If we let $w = e^{i\theta} z$, for any angle θ, then w satisfies the same differential equation as z; that is, $w' = w(e^{i\phi} + w\bar{w})$. You may show this by hand or by using *Maple* (but it's easier to do it by hand).

(b) Plot the trajectories of the differential equation for parameter values $\phi = 0, \pi/4, \pi/3, 2\pi/3$. Determine the direction of the trajectories by evaluating the right-hand side at various values of z. (We suggest plotting the trajectories on the time interval $[-3, 2]$, though for $\phi = 0$ the trajectories diverge rapidly and thus you may have to reduce the right endpoint of this interval in order to avoid an error message.) What kind of symmetries can you see in the plots?

(c) Examine the results of (b). Can you detect whether the parameter has passed through a bifurcation value? What exactly is the qualitative change in the behavior of the solutions?

(d) Plot the trajectories of the system for the two parameter values $\phi = \pi/2 \pm 0.2$. (You should also change the time interval to $[-5, 3]$.) Estimate the bifurcation value of the parameter.

(e) Let $\rho = z\bar{z}$. Note that ρ is the square of the distance from the origin to z. In particular, ρ is a real-valued positive function of t. Moreover, one can show that ρ satisfies the differential equation $\rho' = 2\rho(\operatorname{Re}(e^{i\phi}) + \rho)$. Note that a critical point of this differential equation corresponds to a solution of (i) that stays a fixed distance from the origin. By finding the critical points of the differential equation for ρ, explain the results of (c) and (d).

17. Consider the complex-valued differential equation

$$z' = e^{i\phi} z + \bar{z}^2. \tag{i}$$

Refer to the text preceding Problem 16 for general remarks concerning complex-valued differential equations.

(a) Show that equation (i) is rotationally invariant through an angle of $2\pi/3$ in the following sense: If we let $w = e^{2\pi i/3} z$, then w also satisfies the differential equation (i), i.e., $w' = e^{i\phi} w + \bar{w}^2$. You may show this by hand or by using *Maple*.

(b) Plot the trajectories of the differential equation for parameter values $\phi = 0, \pi/4, \pi/2$. You should use a set of initial conditions consisting of 12 evenly spaced points on each of the circles of radius 0.5 and 1.5. Determine the direction of the trajectories by evaluating the right-hand side at various values of z. (We suggest plotting the trajectories on the time interval $[-3, 2]$.) What symmetries can you detect in the plot? (*Note*: These plots may take several minutes to draw.)

(c) Look at the results of (b). On which interval does the parameter pass through a bifurcation value? What exactly is the qualitative change in the behavior of the solutions?

(d) Set $z = x + iy$ in equation (*i*), expand the equation in terms of x and y, and take the real and imaginary parts to obtain an equivalent real-valued 2×2 system.

(e) Using the result of (d), for each of the parameter values $0, \pi/4, \pi/2$, find the critical points of the system (there are 4 in each case). Indicate these critical points on your plots in (b).

18. Consider the complex-valued differential equation

$$z' = 0.3e^{i\phi}z - (1 + i)z^3\bar{z}^2 + \bar{z}^4. \tag{i}$$

Refer to the text preceding Problem 16 for general remarks concerning complex-valued differential equations.

(a) Show that equation (*i*) is rotationally invariant through an angle of $2\pi/5$ in the following sense: If we let $w = e^{2\pi i/5}z$, then w also satisfies the differential equation (*i*).

(b) Plot the trajectories of the differential equation for parameter values $\phi = 0, \pi/4, \pi/2$. You should use a set of initial conditions consisting of 20 evenly spaced points on each of the circles of radius 0.25 and 0.75. Determine the direction of the trajectories by plotting the vector field associated with this equation. (We suggest plotting the trajectories on the time interval $[-0.3, 10]$, with the range restricted to $[-1, 1]$.) What symmetries can you detect in the plots? (*Note*: These plots may take several minutes to draw.)

(c) Examine the results of (b). On which interval or intervals does the parameter pass through a bifurcation value? What exactly is the qualitative change in the behavior of the solutions? Using the **display** command, plot the vector fields and integral curves on the same graph for each parameter value. By examining these plots, try to figure out how many critical points there are inside the unit square. (*Note*: The stable critical points are easy to find. The unstable ones are not so obvious.)

Glossary

We list here the *Maple* objects that arise in the various input statements encountered in this book. We list them according to five types: commands, options, built-in functions, constants, and programming statements. The distinction between the first and third is somewhat artificial, as *Maple* makes no distinction between them. However, for the purposes of this course, it is convenient to think of a *Maple* built-in function as we normally think of functions, in particular as something that can be evaluated or plotted; while a command is something that manipulates data or expressions, or initiates a process.

Each command and option is followed by a brief description of its effect, and then one or more examples. To find a full description of a command you can type **?command** in a Worksheet.

Maple Commands

@ (at sign)   Composition operator.
   (cos@arccos)(x);

@@ (double at sign)   Repeated composition operator.
   (D@@3)(sin);

conjugate   Gives the complex conjugate of a complex number.
   conjugate(2 + 3*I);

contourplot (in the **plots** package)   Plots the level curves of a function of two variables.
   with(plots):
   contourplot(x^2 + y^2, x = −2..2, y = −2..2);

convert   Converts an expression to a specified form.
   convert(0.425, rational);
   convert(series(sin(x), x), polynom);

D   Differentiation operator for functions; used to describe initial conditions on derivatives.

D(sin);
dsolve({diff(y(x), x$2) = y(x), y(0) = 1, D(y)(0) = 1}, y(x));

DEplot (in the **DEtools** package)   Plots one or more numerical solutions of a differential equation or system of differential equations.

with(DEtools):
DEplot(diff(y(x), x$2) = x^3*y(x) + 1, y(x), x = −3..2, {[y(0) = 0.5, D(y)(0) = 2]}, linecolor = black);
DEplot({diff(x(t), t) = 2*x(t) − y(t), diff(y(t), t) = x(t) + 2*y(t)}, [x(t), y(t)], t = −1..1, {[x(0) = 0.5, y(0) = 2], [x(0) = 1, y(0) = 0], [x(0) = 0, y(0) = 1], [x(0) = 1, y(0) = 1], [x(0) = 0, y(0) = 2]}, x = −10..10, y = −10..10, linecolor = black);

dfieldplot (in the **DEtools** package)   Plots the direction field of a differential equation or system of equations.

with(DEtools):
dfieldplot(diff(x(t), t) = exp(−t) − 2*x(t), x(t), t = −2..3, x = −1..2);
dfieldplot(diff(y(x), x) = x − 3*y(x), y(x), x = −2..2, y = −2..2, axes = BOXED, arrows = LINE);
dfieldplot({diff(x(t), t) = x(t)*(2 − y(t)), diff(y(t), t) = y(t)*sin(x(t))}, [x(t), y(t)], t = 0..1, x = −3..3, y = −3..3, axes = BOXED, arrows = THICK);

diff   Differentiation operator.

diff(f(x), x);
diff(x^3 − 3*x, x);
diff(cos(x), x$2);

display   Combines and displays several previously generated graphics.

display({graph1, graph2});

$ (dollar sign)   Sequence operator.

[j^2 $ j = 0..5];
diff(sinh(x), x$2);

dsolve   Symbolic ODE solver.

dsolve(diff(y(x), x$2) − x*y(x) = 0, y(x));
dsolve({diff(y(x), x) + y(x)^2 = 0, y(0) = 1}, y(x));
dsolve({diff(x(t), t) = 2*x(t) + y(t), diff(y(t), t) = −x(t)}, {x(t), y(t)});

dsolve(. . ., method = laplace)   Computes symbolic solutions of differential equations using the Laplace Transform.   Useful for differential equations involving discontinuous functions.

dsolve({diff(y(x), x) + y(x) = Dirac(x − 1), y(0) = 1}, y(x), method = laplace);

dsolve(..., numeric)   Numerical ODE solver.
   **dsolve({diff(y(x), x) = x + y(x)^2, y(0) = 1}, y(x), numeric, maxfun = 1000,
      startinit = TRUE);**
   **dsolve({diff(y(x), x$2) − x*y(x) = 0, y(0) = 1, D(y)(0) = 0}, y(x), numeric,
      maxfun = 1000, startinit = TRUE);**

dsolve(..., type = series)   Computes series solutions of differential equations.
   dsolve({diff(y(x), x$2) = x*y(x)^2, y(0) = 1, D(y)(0) = 0}, y(x), type = series);

eigenvects (in the **linalg** package)   Computes eigenvalues and eigenvectors of a
matrix.
   with(linalg):
   eigenvects([[a, b], [c, d]]);
   eigenvects([[1, 0, 0], [1, 1, 1], [1, 2, 4]]);

evalc   Evaluates an expression containing complex numbers, assuming that unknown variables represent real numbers.
   evalc(Re(x + I*y));
   evalc(abs(x + I*y));

evalf   Evaluates numerically; gives a decimal approximation of a number.
   evalf(Pi, 15);
   evalf(solve(x^2 − 4*x + 9 = 0, x));

evalf(Int())   Numerical integration.
   evalf(Int(f(x), x = 0..5));
   evalf(Int(sin(x)/x, x = 0..infinity));

evalm   Evaluates a matrix expression. Often used with the matrix multiplication
operator **&***.
   evalm([[a, b], [c, d]] &* [u, v]);

expand   Expands products and powers in an algebraic expression.
   expand((x + y)^3);

factor   Factors a polynomial.
   factor(x^4 − y^4);

fieldplot (in the **plots** package)   Draws a vector field plot given a pair of functions
of two variables; lengths of vectors are proportional to their magnitudes.
   with(plots):
   fieldplot([y, −x], x = −1..1, y = −1..1);
   fieldplot([f(x, y), g(x, y)], x = 0..2, y = 0..5, arrows = SLIM, axes = BOXED);

fsolve   Finds a single numerical solution to an equation or system of equations.
If an interval is specified, looks for a solution in that interval.
   fsolve(cos(x) = x, x, 0..1);
   fsolve({x*y = 2, y = x}, {x, y});

implicitplot (in the **plots** package)   Plots a curve defined implicitly.
  with(plots):
  implicitplot(y^2 + y = x^3 − x, x = −2..3, y = −3..3);

int   Integration operator for both definite and indefinite integrals.
  int(1/(1 + x^2), x);
  int(exp(−x), x = 0..infinity);

invlaplace (in the **inttrans** package)   Computes the inverse Laplace Transform.
  with(inttrans):
  invlaplace(exp(−s)/s, s, t);

laplace (in the **inttrans** package)   Computes the Laplace Transform.
  with(inttrans):
  laplace(sin(t), t, s);

limit   Finds a limit.
  limit(sin(x)/x, x = 0);
  limit(1/x, x = 0, left);

linsolve (in the **linalg** package)   Solves a matrix equation.
  with(linalg):
  linsolve([[1, 2], [3, 4]], [6, 7]);

map   Applies a function to each element of a list or set.
  map(sqrt, {1, 2, 3});
  map(x −> x^2, [1, 2, 3]);

matrix (in the **linalg** package)   Converts a list into a matrix, and displays it accordingly.
  with(linalg):
  matrix([[1, 2, 3], [4, 5, 6]]);

odeplot (in the **plots** package)   Plots a numerical solution produced by dsolve(..., numeric).
  with(plots):
  sol := dsolve({diff(y(x), x) = x^2 − y(x), y(0) = 1}, y(x), numeric, maxfun = 1000);
  odeplot(sol, [x, y(x)], 0..2);

op   Extracts elements of a list, set, sequence or expression. Can be used to remove extra levels of brackets or braces.
  op([[1, 2], 3, [4, 5, 6, [7]]]);
  map(op, [[1, 2], 3, [4, 5, 6, [7]]]);

plot   Basic plot command. Plots functions of one variable, expressions, lists of points, parametric curves, or sets containing any of the previous objects.

    plot(sin, 0..Pi);
    plot(f(x), x = 0..1);
    plot({x^2, x^3}, x = −2..2);
    plot([[1, 1], [2, 4], [3, 10]]);
    plot([cos(t), sin(t), t = 0..2*Pi]);
    plot({f(t, j) \$ j = 1..10}, t = 0..5);

print   Shows the value of an expression; useful for displaying lists as tables.

    map(print, [1, 2, 3, 4]):

rhs   Gives the right-hand side of an equation.

    rhs(y = 2*x);
    rhs(dsolve(diff(y(x), x) = x*y(x), y(x)));

seq   Builds a sequence; useful for constructing lists of initial conditions.

    seq(x^2 + c, c = −3..3);
    seq(i*(i + 1), i = [1, 2, 3, 5, 7]);
    seq(seq([i, j], i = 1..5), j = 1..5);

series   Generates terms of a power series expansion about a point.

    series(exp(−x), x = 0, 10);

simplify   Simplifies an expression.

    simplify(1/(1 + x)^2 − 1/(1 − x)^2);

solve   Solves an equation or set of equations.

    solve(2*x^2 − 3*x + 6 = 0, x);
    solve({x + 3*y = 4, −x − 5*y = 3}, {x, y});

subs   Substitutes for parts of an expression.

    subs(x = 2, x^3 − 4*x + 1);
    subs(sin(x) = z, sin(x)^2 + cos(x));
    subs({x(t) = t^2, y(t) = cos(t)}, {x(t), y(t)});

sum   Sums a sequence.

    sum((i − 1)/(i + 1), i = 1..10);
    sum(g(0.1*j), j = 20..30);

textplot (in the **plots** package)   Plots text strings (for labeling).

    with(plots):
    textplot([1, 1, `The point (1, 1)`]);

unapply   Converts an expression into a function.

    f := unapply(x^2 + 3, x);
    f := unapply(x^2 + k, x, k);

with   Loads a *Maple* package.
 with(plots):
 with(DEtools):
 with(linalg):
 with(inttrans):

Global Variables and Options to Maple Commands

adaptive   Disables adaptive feature of the **plot** command.
 plot(sin(x^100), x = 0..1, adaptive = false);

arrows   Controls the shape of the arrows in a vector field or direction field.
 dfieldplot(diff(x(t), t) = exp(−t) − 2*x(t), x(t), t = −2..3, x = −1..2, arrows =
  LINE);
 DEplot(diff(y(x), x) = x*y(x) + 2, y(x), x = 0..2, {[y(0) = 0]}, y = 0..5, arrows =
  SLIM, linecolor = black);

axes   Controls how the axes of a plot are displayed.
 plot(sin(x), x = 0..2*Pi, axes = FRAMED);
 dfieldplot(diff(x(t), t) = exp(−t) − 2*x(t), x(t), t = −2..3, x = −1..2, axes = BOXED);

color   Specifies the color of curves drawn by **plot** and arrows drawn by **dfieldplot**
and **DEplot**; default is red.
 plot({sin(x), cos(x)}, x = 0..Pi, color = black);

contours   Specifies how many contours to use in a contour plot.
 contourplot(f(x, y), x = 0..1, y = 0..2, contours = 12);

Digits   This global variable controls the number of digits used in numerical
computations; default is 10.
 Digits := 20;

dirgrid   Controls the number of arrows plotted by **dfieldplot** and **DEplot**.
 dfieldplot(diff(y(x), x) = x*y(x)^3 − 2, y(x), x = 0..Pi, y = 0..1, dirgrid = [10, 10]);

grid   Controls the number of points plotted by **contourplot** and **implicitplot**;
default is 25 × 25.
 contourplot(x^2 − y^3, x = −2..2, y = −2..2, grid = [50, 50]);

linecolor   Specifies the color of solution curves drawn by **DEplot**; default is
yellow.
 DEplot({diff(x(t), t) = y(t), diff(y(t), t) = −x(t)}, [x(t), y(t)], t = −1..1, {[x(0) = 1,
  y(0) = 0]}, linecolor = black):

maxfun   Limits the number of steps used by **dsolve(..., numeric)**; default is
30, 000. Also applies to **DEplot** with **method = rk45**.
 dsolve({diff(y(x), x) = x^2 − y(x)^2, y(0) = 1}, y(x), numeric, maxfun = 1000);

method   Specifies a method in **dsolve**; possibilities include **classical[foreuler]**, **gear**, and **lsode**.

   dsolve({diff(y(x), x) = sin(x^100), y(0) = 0.5}, y(x), numeric, method = gear);

numpoints   The minimum number of points used by a plotting routine; default is 49.

   plot(sin(Pi*x), x = 0..48, adaptive = false, numpoints = 100);

obsrange   Indicates if *Maple* should halt once the solution curve has passed outside of the specified range; default is TRUE.

   DEplot(diff(y(x), x$2) = x^3*y(x) + 1, y(x), x = −3..2, {[y(0) = 0.5, D(y)(0) = 2]}, y = −1..2, obsrange = FALSE);

Order   This global variable controls the order of series calculations; default is 6.

   Order := 11;

sample   Supplies a set of points to be used in the **plot** command; especially when adaptive plotting has been disabled.

   plot(f, adaptive = false, sample = [seq(0.1*i, i = 1..50)]);

scaling   Controls whether the axes in a plot have the same scale. Default is **UNCONSTRAINED**, so axes may have different scales.

   plot([2*cos(t), 3*sin(t), t = 0..2*Pi], scaling = CONSTRAINED);

scene   Controls which variables are plotted. Used in the plotting commands from the **DEtools** package.

   DEplot({diff(x(t), t) = −t*cos(y(t)), diff(y(t), t) = cos(x(t))}, [x(t), y(t)], t = 0..10, {[x(0) = 1, y(0) = 1]}, x = 0..3, y = 1..3, scene = [t, y], stepsize = 0.1);

stepsize   Specifies the number of points used by **DEplot** to draw a solution curve; default is 1/20 the size of the range of the independent variable.

   DEplot(diff(y(x), x$2) = x^3*y(x) + 1, y(x), x = −3..2, {[y(0) = 1, D(y)(0) = 2]}, stepsize = 0.1);

title   Specifies a title for a graph.

   plot(sin(x), x = 0..4*Pi, title = `Another sine wave`);

type   Specifies a symbolic method for **dsolve**.

   dsolve({diff(y(x), x$2) = y(x)^2 + x^2, y(0) = 1, D(y)(0) = 0}, y(x), type = series);

Built-In Functions

abs $|x|$.

AiryAi   The solution Ai(x) to Airy's equation $y'' - xy = 0$.

AiryBi   The solution Bi(x) to Airy's equation.

arccos   arccos x.

arcsin   arcsin x.

BesselJ(0, ·)   The solution $J_0(x)$ to Bessel's equation of order zero, $x^2 y'' + xy' + x^2 y = 0$.

BesselY(0, ·)   The solution $Y_0(x)$ to Bessel's equation of order zero.

cos $\cos x$.

cosh $\cosh x$.

Dirac   The Dirac delta function.

erf   The *error function* $\mathrm{erf}(x) = (2/\sqrt{\pi}) \int_0^x e^{-t^2} \, dt$.

exp e^x.

Heaviside   The unit step function.

Im $\Im(z)$, the imaginary part of a complex number.

ln   The natural logarithm $\ln x = \log_e x$.

Re $\Re(z)$, the real part of a complex number.

Si   The *sine integral* $\mathrm{Si}(x) = \int_0^x \sin(t)/t \, dt$.

sin $\sin x$.

sinh $\sinh x$.

sqrt $\sqrt{x}$.

tan $\tan x$.

tanh $\tanh x$.

Built-In Constants

I $i = \sqrt{-1}$.

infinity ∞.

Pi π.

Maple Programming

do . . . od   Marks a block of statements for repetition.
   j := 10; while j > 0 do j := j − 1; od;

for   Repeats a command a certain number of times.
   a := [ ]; for j from 0 to 1 by 0.2 do a := [op(a), [j, j^2]]; od;

global   Specifies that certain variables in a program are global, *i.e.*, their values may come from outside the program.

if . . . fi   Introduces a conditional.
   checksign := x –> if x < 0 then print(negative); else print(nonnegative); fi;

local   Specifies that certain variables in a program are local, *i.e.,* are only used inside the program.

proc ... end   Defines a *Maple* procedure or program.

```
eqn := diff(y(x), x) = x^2 − y(x);

compareplots := proc(b) local exact, approx; global eqn;
   exact := plot(rhs(dsolve({eqn, y(0) = b}, y(x))), x = 0..5, style = POINT):
   approx := odeplot(dsolve({eqn, y(0) = b}, y(x), numeric,
        maxfun = 1000), [x, y(x)], 0..5):
   display({exact, approx});
end;

compareplots(1);
```

Sample Worksheet Solutions

Problem Set B, Problem 1.

We are interested in the differential equation $x\left(\frac{\partial}{\partial x}y(x)\right) + 2\,y(x) = \sin(x)$, with initial condition $y(\frac{\pi}{2}) = c$. Here is the general solution:

```
>    restart:
```

```
>    ivp := {x*diff(y(x), x) + 2*y(x) = sin(x), y(Pi/2) = c};
```

$$ivp := \left\{x\left(\frac{\partial}{\partial x}y(x)\right) + 2\,y(x) = \sin(x),\ y\!\left(\frac{1}{2}\pi\right) = c\right\}$$

```
>    sol := dsolve(ivp, y(x));
```

$$sol := y(x) = -\frac{-\sin(x) + x\cos(x) + 1 - \frac{1}{4}\,c\,\pi^2}{x^2}$$

Here is a function that gives, for fixed c, the solution of the IVP with initial condition $y(\frac{\pi}{2}) = c$.

```
>    f := unapply(simplify(rhs(sol)), c);
```

$$f := c \to \frac{1}{4}\,\frac{4\sin(x) - 4\,x\cos(x) - 4 + c\,\pi^2}{x^2}$$

The solution with $y(\frac{\pi}{2}) = 1$ is given by:

```
>    y1 := unapply(f(1), x);
```

$$y1 := x \to \frac{1}{4}\,\frac{4\sin(x) - 4\,x\cos(x) - 4 + \pi^2}{x^2}$$

(a)

Here is a table of values for the solution of the IVP with $c = 1$.

```
>    pnt := x -> [x, evalf(y1(x))]:
```

```
>    list_of_points := [pnt(0.5*x) $ x = 1..10]:
```

```
>    map(print, list_of_points):
```

$$[.5,\ 6.032141434]$$

$$[1.0, 1.768569780]$$

$$[1.5, 1.048351238]$$

$$[2.0, .8022480500]$$

$$[2.5, .6509971652]$$

$$[3.0, .5087220664]$$

$$[3.5, .3587115329]$$

$$[4.0, .2078233180]$$

$$[4.5, .07103467064]$$

$$[5.0, -.03639336405]$$

(b)

We graph the solution $y_1(x)$ on several intervals.

```
>  plot(y1(x), x = 0..2);
```

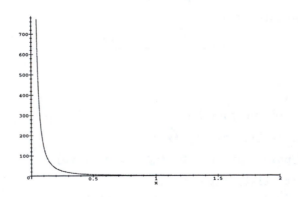

```
>  plot(y1(x), x = 1..10);
```

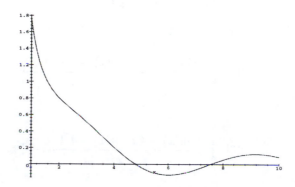

```
>   plot(y1(x), x = 10..100);
```

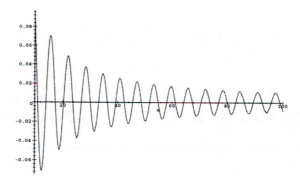

By inspecting the function $y_1(x)$, we see that the solution approaches positive infinity as x goes to $x = 0$, and approaches $x = 0$ as x goes to ∞. This analysis is borne out by the graphs above.

(c)

Here is a table of the solutions of the IVP with initial conditions $y(\frac{\pi}{2}) = .2\,j$ for $j = 1, ..., 5$, and a plot of these solutions over two different intervals. Note that the index $j = 1$ corresponds to the lowest solution, and the index $j = 5$ corresponds to the highest.

```
>   map(print, [[[j], f(0.2*j)] $ j = 1..5]):
```

$$[[1], \ \frac{1}{4} \ \frac{4\sin(x) - 4\,x\cos(x) - 4 + .2\,\pi^2}{x^2}]$$

$$[[2], \ \frac{1}{4} \ \frac{4\sin(x) - 4\,x\cos(x) - 4 + .4\,\pi^2}{x^2}]$$

$$[[3], \ \frac{1}{4} \ \frac{4\sin(x) - 4\,x\cos(x) - 4 + .6\,\pi^2}{x^2}]$$

$$[[4], \ \frac{1}{4} \ \frac{4\sin(x) - 4\,x\cos(x) - 4 + .8\,\pi^2}{x^2}]$$

$$[[5], \ \frac{1}{4} \ \frac{4\sin(x) - 4\,x\cos(x) - 4 + 1.0\,\pi^2}{x^2}]$$

```
> plot({f(0.2*j) $ j = 1..5}, x = 0.1..0.2, color = black);
```

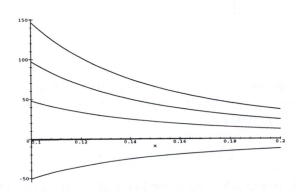

```
> plot({f(0.2*j) $ j = 1..5}, x = 0.2..10, y = -1..1,
color = black);
```

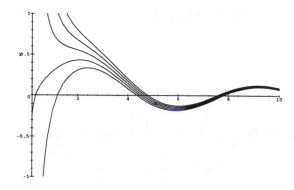

(d)

The solutions in (c) approach plus or minus infinity as x approaches 0 from the right, and approach 0 as x approaches infinity. Can we find a solution that has no singularity at $x = 0$? The general solution is given at the beginning of this problem. The only way that a solution could possibly be nonsingular at $x = 0$ is if $c = \frac{4}{\pi^2}$. We can check using L'Hopital's rule that in this case the function is continuous at 0. Here is a graph of this solution.

```
>   plot(f(4/Pi^2), x = -5..5);
```

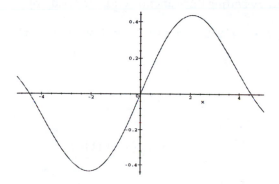

Problem Set B, Problem 4.

We are interested in solutions to $\frac{\partial}{\partial x} y = \frac{x - e^{(-x)}}{y + e^y}$, particularly the solution with $y(1.5) = .5$.

(a)

```
>   restart:
```

```
>   eqn := diff(y(x), x) = (x - exp(-x))/(y + exp(y));
```

$$eqn := \frac{\partial}{\partial x} y(x) = \frac{x - e^{(-x)}}{y + e^y}$$

```
>   dsolve(eqn, y(x));
```

$$\frac{1}{2} y(x)^2 + e^{y(x)} - \frac{1}{2} x^2 - e^{(-x)} = \_C1$$

(b)

The solutions of this equation are implicit; to use them, we define a function $f(x, y)$ equal to the left hand side of the equation reported by Maple. The solution of the IVP is obtained by setting this function equal to f(1.5, .5).

```
>   f := (x,y) -> y^2/2 + exp(y) - x^2/2 - exp(-x);
```

$$f := (x, y) \rightarrow \frac{1}{2} y^2 + e^y - \frac{1}{2} x^2 - e^{(-x)}$$

```
>   c := f(1.5, 0.5);
```

$$c := .4255911109$$

We can use contourplot to see some of the solution curves.

```
>   with(plots):
```

```
>   contourplot(f(x, y), x = -1..3, y = -2..2, axes = BOXED,
color = black);
```

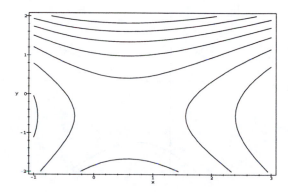

(c), (d)

Here are the values of the solution of the IVP at the points $x = 0, 1, 1.8$, and 2.1.

```
> fsolve(f(0, y) = c);
```

$$.3183856549$$

```
> fsolve(f(1, y) = c);
```

$$.2356325919$$

```
> fsolve(f(1.8, y) = c);
```

$$.6821874213$$

```
> fsolve(f(2.1, y) = c);
```

$$.8662118724$$

The following graph shows the solution corresponding to the value f(1.5, .5). The top curve corresponds to the solution of the IVP because it passes through the point (1.5, .5). We have marked the points with x-coordinates 0, 1, 1.8, and 2.1.

```
> curve := implicitplot(f(x, y) - c, x = -1..3, y = -2..2,
color = black):
```

```
> points := plot([[0, 0.318], [1, 0.236], [1.8, 0.682],
[2.1, 0.866]], style = POINT, symbol = BOX):
```

```
> display({curve, points});
```

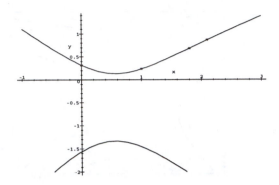

Problem Set C, Problem 3.

We are going to investigate the initial value problem y' = (y-x)(1-y^3), y(0) = b for various choices of the initial value b. We first try to find an exact solution with dsolve.

```
> restart;
> de := diff(y(x), x) = (y(x) - x)*(1 - y(x)^3);
```

$$de := \frac{\partial}{\partial x} y(x) = (y(x) - x)(1 - y(x)^3)$$

```
> dsolve(de, y(x));
```

The lack of a response here shows that dsolve cannot find a solution.

(a)

We will use DEplot to graph the solutions of the IVP for the initial conditions y(0) = -0.5, 0, 0.5, 1.0, 1.5, 2.0.

```
> with(DEtools):
> iniset := {[y(0) = -0.5], [y(0) = 0.0], [y(0) = 0.5],
[y(0) = 1.0], [y(0) = 1.5], [y(0) = 2.0]}:
> DEplot(de, y(x), x = 0..2.0, iniset, method = rkf45,
arrows = NONE, linecolor = black);

Error, (in DEtools/DEplot/drawlines) Stopping integration due to,
rkf45 is unable to achieve requested accuracy
```

Perhaps some of the graphs have features, e.g., asymptotes, that lead to the failure to achieve the requested accuracy. Let's plot over a shorter interval in hope of revealing the problem.

```
>  DEplot(de, y(x), x = 0..0.5, iniset, method = rkf45,
arrows = NONE, linecolor = black);
```

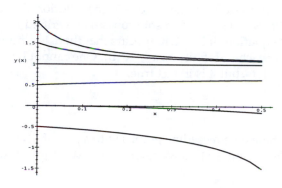

It appears that solutions with initial values $b < 0$ approach negative infinity at finite x-values, and it is likely that it is difficult to achieve accuracy when computing such solutions. We now plot over a longer interval, but with the y-range restriction y = -1..5. We also use the option stepsize = 0.1 to get a better (smoother) graph.

```
>  plot1 := DEplot(de, y(x), x=0..5, iniset, method = rkf45,
arrows = NONE, y = -1..5, linecolor = black, stepsize = 0.1):

>  plot1;
```

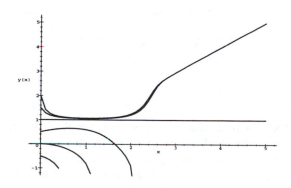

This plot suggests that solutions with initial values b < 0, in fact with initial values b < 1, approach -infinity at finite x-values.

(b)

The plots in part (a) indicate that if b = 1, the corresponding solution is y = 1; if 0 < b < 1, the solution first increases, and then decreases to -infinity; if b <= 0, the solution decreases to -infinity; and if b > 1, the solution first decreases and then increases to infinity. That y = 1 is a solution can be verified by plugging y = 1 into the differential equation. If b < 1, it appears that there is a number x*(b) such that the solution approaches -infinity as x -> x*(b). One cannot be certain about this on the basis of the plots, but it is in fact true.

(c)

Now we'll combine the preceeding graphs with a plot of y = x. We'll plot y = x with the option style = point, so that we can distinguish its graph from the preceeding graphs.

```
>   line := plot(x, x = 0..5, style = POINT):

>   with(plots):

>   display({plot1, line});
```

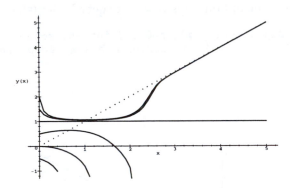

So it appears that for b>1, the solutions are asymptotic to the line y = x. To understand this on the basis of the differential equation, note that along the line y = x, y' is zero; above the line, y' is negative; and below the line, y' is positive. Thus the solutions are pushed toward the line y = x. Here is a plot of the direction field together with several solutions and the line y = x. We see that the direction field confirms the observations we have made about the solutions.

```
>  plot2 := dfieldplot(de, y(x), x = 0..5, y = -1..5,
arrows = LINE, axes = BOXED):
```

```
>  display({plot1, plot2, line});
```

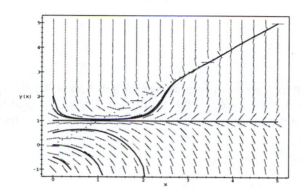

We can also plot the direction field and the numerical solutions at the same time with the DEplot command.

```
>  plot3 := DEplot(de, y(x), x = 0..5, iniset,
method = rkf45, y = -1..5, linecolor = black, stepsize = 0.1):
```

```
>  display({plot3, line});
```

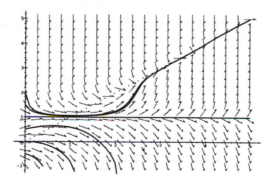

Problem Set D, Problem 1.

```
>  restart:
```

We consider the second order differential equation

```
>  airy := diff(y(x), x$2) = x*y(x);
```

$$airy := \frac{\partial^2}{\partial x^2} y(x) = x\, y(x)$$

(a)

When x is close to 0, we want to compare the solution of Airy's equation with the solution of the IVP y"(x) = 0, y'(0) = 1, y(0) = 0. The solution of this IVP is y = x. Here is a plot of the numerical solution of Airy's equation together with the function y = x.

```
>  with(plots):  with(DEtools):
```

```
>  fac1plot := plot(x, x = -2..2):
```

```
>  airy1plot := DEplot(airy, y(x), x = -2..2, {[y(0) = 0,
D(y)(0) = 1]}, method = rkf45, linecolor = black):
```

```
>  display({fac1plot, airy1plot});
```

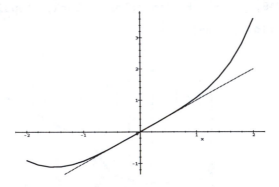

The facsimile solution agrees well with the actual solution in a neighborhood of 0.

(b)

For x close to $-16 = -4^2$, we want to compare the solution of Airy's equation to the facsimile solution

```
>  fac2 := (c1, c2) -> c1*sin(4*x + c2);
```

$$fac2 := (c1, c2) \rightarrow c1 \sin(4x + c2)$$

First we'll plot a numerical solution of Airy's equation.

```
>  airy2plot := DEplot(airy, y(x), -18..-14, {[y(0) = 0,
D(y)(0) = 1]}, method = rkf45, stepsize = 0.1, linecolor = black):
```

```
>  airy2plot;
```

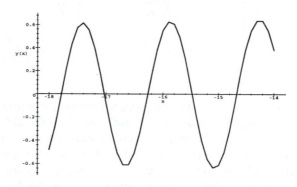

This certainly looks like a sine wave. Let's see how well it matches up with an appropriate sine wave. We have to choose constants c_1 and c_2 in the facsimile solution to make it match up. Note first that c_1 is the amplitude of the facsimile solution, and we can see from the graph that the amplitude of the solution of Airy's equation is about .61 on this interval. The constant c_2 determines the phase shift, and can be read off from the zeros of the solution. In particular, the solution of Airy's equation has a zero at about -16.3, so we should choose c_2 to satisfy $4(-16.3) + c_2 = 0$.

```
>  fac2plot := plot(fac2(0.61, 4*16.3), x = -18..-14,
style = POINT):
```

```
>  display({fac2plot, airy2plot});
```

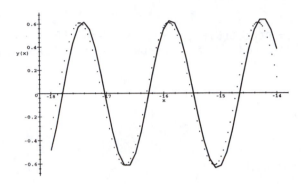

The plots match up well. Why did we have to choose the values of c_1 and c_2 by hand? Our analysis suggests that the facsimile solution should approximate the actual solution near $x = -K^2$. But the initial condition that we used for Airy's equation is at the point $x = 0$, where the facsimile solution is not a good approximation. Thus, the undetermined constants in the facsimile solution are unrelated to the initial conditions in Airy's equation, and we needed to choose them by hand to get a good match. Nevertheless, the frequency of the facsimile solution is determined by K, and is independent of the undetermined constants. Thus, we can at least conclude that the frequency of the solutions of Airy's equation in a neighborhood of $x = -K^2$ will be almost proportional to K.

(c)

```
>  fac3 := (c1, c2) -> c1*sinh(4*x + c2);
```

$$fac3 := (c1,\, c2) \rightarrow c1 \sinh(4\,x + c2)$$

```
>  airy3plot := DEplot(airy, y(x), 14..18, {[y(0) = 0,
D(y)(0) = 1]}, method = rkf45, linecolor = black):
```

```
>  airy3plot;
```

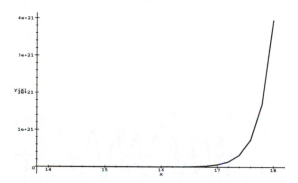

We'd like to compare this with the graph of the hyperbolic sine function. Again, we have to choose c_1 and c_2 by hand. We'll start with an arbitrary choice: $c_1 = 1$, $c_2 = 0$.

```
>  plot(fac3(1, 0), x = 14..18);
```

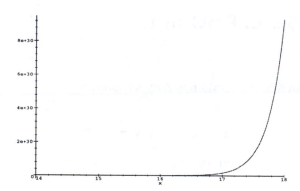

The graphs are remarkably similar. Note, however, that the values in the second graph are about 10^9 greater than in the first, which means that we should have chosen c_1 to be about $10^{(-9)}$.

(d)

Finally, we want to produce a graph of the numerical solution on the interval from -20 to 2.

```
>  DEplot(airy, y(x), -20..2, {[y(0) = 0,
D(y)(0) = 1]}, method = rkf45, stepsize = 0.1, linecolor = black);
```

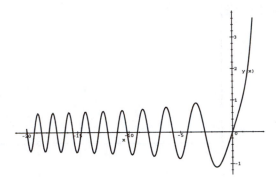

This graph indicates that the frequency increases and the amplitude decreases as x goes to $-\infty$. The increasing frequency is predicted by our facsimile analysis; the decreasing amplitude is harder to explain.

Problem Set E, Problem 1.

(a)

We are interested in the solution of Airy's equation.

```
>  restart:
```

```
>  airyeqn := diff(y(x), x$2) - x*y(x) = 0;
```

$$airyeqn := (\frac{\partial^2}{\partial x^2} y(x)) - x\, y(x) = 0$$

```
>  Order := 11:
```

```
>  seriessol := dsolve({airyeqn, y(0) = 0, D(y)(0) = 1}, y(x),
type = series);
```

$$seriessol := y(x) = x + \frac{1}{12} x^4 + \frac{1}{504} x^7 + \frac{1}{45360} x^{10} + O(x^{11})$$

```
>  taylorsol := convert(rhs(seriessol), polynom);
```

$$taylorsol := x + \frac{1}{12} x^4 + \frac{1}{504} x^7 + \frac{1}{45360} x^{10}$$

(b)

Here is the exact solution of Airy's equation.

```
> exactsol := dsolve({airyeqn, y(0) = 0, D(y)(0) = 1}, y(x));
```

$$exactsol := y(x) = \frac{2}{9}\,\frac{\pi\,3^{5/6}\,\sqrt{x}\,\text{BesselI}(\frac{1}{3},\,\frac{2}{3}\,x^{3/2})}{\Gamma(\frac{2}{3})}$$

Here is a plot comparing the exact solution to the (truncated) series solution.

```
> plot1 := plot(taylorsol, x = 0..6, linestyle = 3):
> plot2 := plot(rhs(exactsol), x = 0..6):
> with(plots):
> display({plot1, plot2});
```

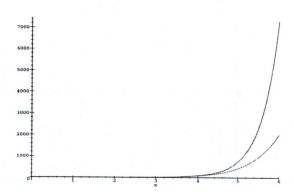

(c)

The general solution of the comparison equation $y' = y$ is $c_1\,e^x + c_2\,e^{(-x)}$. The solution blows up exponentially (when c_1 is nonzero). Thus the exact solution of Airy's equation will diverge exponentially from any Taylor polynomial approximation.

Problem Set E, Problem 16.

(a)

```
> restart:
```

We define a square wave on the interval $[0, 10\,\pi]$ by using a sum of Heaviside functions.

```
>  h := t -> -1 + 2*sum((-1)^i*Heaviside(t - Pi*i), i = 0..10);
```

$$h := t \rightarrow -1 + 2\left(\sum_{i=0}^{10} (-1)^i \, \text{Heaviside}(t - \pi\, i)\right)$$

```
>  h(t);
```

$$-1 + 2\,\text{Heaviside}(t) - 2\,\text{Heaviside}(t - \pi) + 2\,\text{Heaviside}(t - 2\,\pi)$$
$$- 2\,\text{Heaviside}(t - 3\,\pi) + 2\,\text{Heaviside}(t - 4\,\pi) - 2\,\text{Heaviside}(t - 5\,\pi)$$
$$+ 2\,\text{Heaviside}(t - 6\,\pi) - 2\,\text{Heaviside}(t - 7\,\pi) + 2\,\text{Heaviside}(t - 8\,\pi)$$
$$- 2\,\text{Heaviside}(t - 9\,\pi) + 2\,\text{Heaviside}(t - 10\,\pi)$$

```
>  plot(h(t), t = 0..30);
```

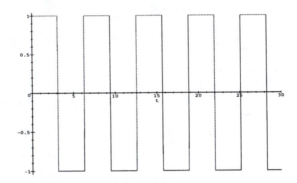

(b)

```
>  eqnlhs := diff(y(t), t$2) + y(t);
```

$$eqnlhs := \left(\frac{\partial^2}{\partial t^2}\, y(t)\right) + y(t)$$

```
>  ic := y(0) = 0, D(y)(0) = 1;
```

$$ic := y(0) = 0,\ D(y)(0) = 1$$

```
>  solb := dsolve({eqnlhs = h(t), ic}, y(t), method = laplace):
>  plot({rhs(solb), h(t)}, t = 0..30);
```

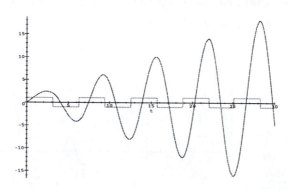

We do see resonance. The solution appears to grow without bound. Let's check the solution by using dsolve(..., numeric).

```
>  numsolb := dsolve({eqnlhs = h(t), ic}, y(t), numeric,
maxfun = 5000);
```

$$numsolb := \mathbf{proc}(rkf45\_x) \ldots \mathbf{end}$$

```
>  yb := u -> subs(numsolb(u), y(t));
```

$$yb := u \rightarrow \mathrm{subs}(\mathrm{numsolb}(u), \mathrm{y}(t))$$

```
>  plot(yb, 0..30, adaptive = false, numpoints = 100);
```

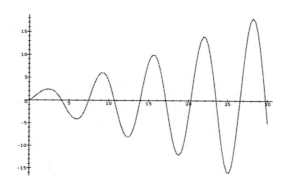

(c)

```
> solc := dsolve({eqnlhs = h(t/2), ic}, y(t), method = laplace):
> plot({rhs(solc), h(t/2)}, t = 0..30);
```

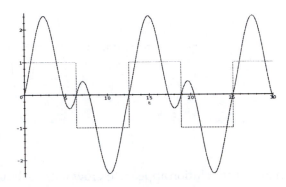

This solution does not exhibit resonance. The solutions remain bounded.

```
> numsolc := dsolve({eqnlhs = h(t/2), ic}, y(t), numeric,
maxfun = 5000);
```

$$numsolc := \mathbf{proc}(rkf45\_x) \ldots \mathbf{end}$$

```
> yc := u -> subs(numsolc(u), y(t)):
> plot(yc, 0..30, adaptive = false, numpoints = 100);
```

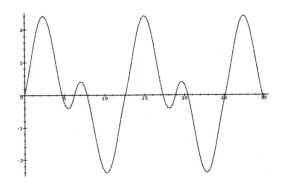

(d)

Now we examine the solution with forcing function h(2 t), which has period π.

```
>  sold := dsolve({eqnlhs = h(2*t), ic}, y(t), method = laplace):
>  plot({rhs(sold), h(2*t)}, t = 0..15);
```

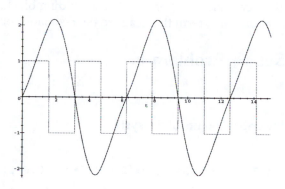

This solution does not exhibit resonance.

```
>  numsold := dsolve({eqnlhs = h(2*t), ic}, y(t), numeric,
maxfun = 5000);
```

$$numsold := \mathbf{proc}(rkf45\_x) \ldots \mathbf{end}$$

```
>  yd := u -> subs(numsold(u), y(t)):
>  plot(yd, 0..15, adaptive = false, numpoints = 100);
```

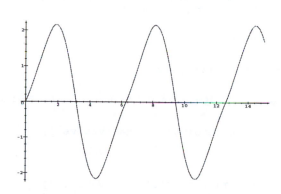

(e)

The results of parts (b), (c), and (d) indicate that resonance may be a consequence of the period of the forcing function, regardless of the actual shape of the function. This is, in fact, true. A rigorous justification of this fact involves the theory of Fourier analysis, which shows that any periodic function can be "built up" out of cosine and sine functions of different periods. Most (but not all!) functions with minimal period 2π contain $\sin(x)$ or $\cos(x)$ as a building block, and it is the occurrence of the $\sin(x)$ or $\cos(x)$ term that leads to the resonance observed in part (b).

Problem Set F, Problem 6.

(a)

First, we find the critical points of the system.

```
> restart:
```

```
> sys := x*(1.5 - x - 0.5*y), y*(2 - 1.5*x - 0.5*y);
```

$$sys := x\,(1.5 - x - .5\,y),\ y\,(2 - 1.5\,x - .5\,y)$$

```
> critpoints := solve({sys}, {x, y});
```

$$critpoints :=$$
$$\{x = 0,\, y = 0\},\, \{y = 0,\, x = 1.500000000\},\, \{x = 0,\, y = 4.\},\, \{y = 1.,\, x = 1.\}$$

We want to compute the eigenvalues of the associated linear system at each critical point. First we define an expression which gives the matrix of first order partial derivatives of the system at a point (x, y).

```
> with(linalg):
derivative := [[diff(sys[1], x), diff(sys[1], y)],
[diff(sys[2], x), diff(sys[2], y)]];
```

Warning, new definition for norm

Warning, new definition for trace

$$derivative := [[1.5 - 2\,x - .5\,y,\ -.5\,x],\ [-1.5\,y,\ 2 - 1.5\,x - 1.0\,y]]$$

```
> eigenvals(subs(critpoints[1], derivative));
```

$$1.500000000,\, 2.$$

```
> eigenvals(subs(critpoints[2], derivative));
```

$$-1.500000000,\, -.2500000000$$

```
>  eigenvals(subs(critpoints[3], derivative));
```
$$-2., -.5000000000$$

```
>  eigenvals(subs(critpoints[4], derivative));
```
$$-1.651387819, .1513878189$$

Thus we find that the types of the four equilibrium points are:
$(0, 0)$ – unstable node
$(1.5, 0)$ – asymptotically stable node
$(0, 4)$ – asymptotically stable node
$(1, 1)$ – unstable saddle point

(b)

```
>  with(DEtools):
```

```
>  tsys := subs({x = x(t),y = y(t)}, [sys]);
```
$$tsys := [x(t) (1.5 - x(t) - .5 y(t)), y(t) (2 - 1.5 x(t) - .5 y(t))]$$

```
>  dsys := diff(x(t), t) = tsys[1], diff(y(t), t) = tsys[2];
```

$dsys :=$
$$\frac{\partial}{\partial t} x(t) = x(t) (1.5 - x(t) - .5 y(t)), \quad \frac{\partial}{\partial t} y(t) = y(t) (2 - 1.5 x(t) - .5 y(t))$$

```
>  vfield := dfieldplot([dsys], [x(t), y(t)], t = -10..10, x = 0..3,
y = 0..5):
vfield;
```

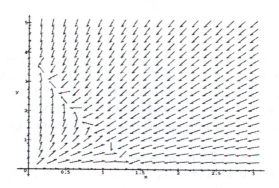

The vector field confirms what we found in part (a).

(c)

We're going to use the DEplot command to plot a phase portrait for this nonlinear system. We start with a rectangular grid of 42 different initial conditions. Note that an initial condition for the system consists of a pair [x(t0) = x0, y(t0) = y0], where t0 is the initial time, and (x0, y0) is the initial position. We take t0 = 0 in each case.

```
>  initials := seq(seq([x(0) = 0.5*i, y(0) = j],
i = 0..6), j = 0..5):
```

```
>  portrait := DEplot([dsys], [x(t), y(t)], t = 0..10, {initials},
x = 0..3, y = 0..5, stepsize = 0.1, method = rkf45,
arrows = NONE, linecolor = black):
portrait;
```

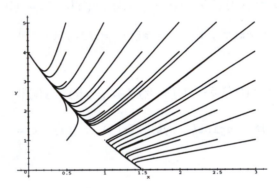

This graph is not ideal. The problem is that a rectangular grid of initial conditions is not the best choice for this system. By looking at the vector field, we see that a better choice of initial conditions would be a collection of evenly spaced points along lines parallel to and above and below the line from (0, 4) to (1.5, 0). Since that line has slope about -2.5, we redefine our initial conditions as follows.

```
>  initials := seq(seq([x(0) = j*(0.1*i),
y(0) = j*(2.5 - 2.5*(.1*i))], i = 0..10), j = [0.6, 2.6]):
```

```
>  portrait := DEplot([dsys], [x(t), y(t)], t = 0..10, {initials},
x = 0..3, y = 0..5, stepsize = 0.1, method = rkf45,
arrows = NONE, linecolor = black):
portrait;
```

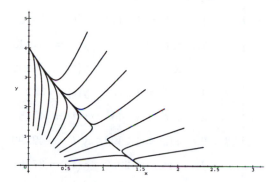

Much better! By superimposing the vector field from part (b), we can get a clear idea of the directions of the trajectories.

```
>  with(plots):
```

```
>  display({vfield, portrait});
```

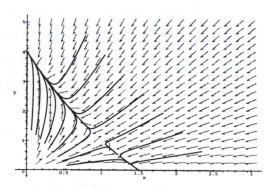

(d)

To find particular values of the population we have to use dsolve(...,numeric). We're interested in the population as a function of t, given initial conditions x0 = 2.5, y0 = 2.

```
>  population := dsolve({dsys, x(0) = 2.5, y(0) = 2}, {x(t), y(t)},
numeric, maxfun = 1000);
```

$$population := \mathbf{proc}(rkf45\_x) \ldots \mathbf{end}$$

Here is a list of values of the population for t = 1..20, for initial conditions (2.5, 2).

```
>   array([seq(evalf(population(u), 5), u = 1..20)]);
```

$$\begin{bmatrix}
t = 1. & x(t) = 1.2944 & y(t) = .76107 \\
t = 2. & x(t) = 1.2026 & y(t) = .62945 \\
t = 3. & x(t) = 1.2035 & y(t) = .56992 \\
t = 4. & x(t) = 1.2221 & y(t) = .52047 \\
t = 5. & x(t) = 1.2450 & y(t) = .47182 \\
t = 6. & x(t) = 1.2692 & y(t) = .42299 \\
t = 7. & x(t) = 1.2938 & y(t) = .37457 \\
t = 8. & x(t) = 1.3181 & y(t) = .32747 \\
t = 9. & x(t) = 1.3416 & y(t) = .28263 \\
t = 10. & x(t) = 1.3638 & y(t) = .24085 \\
t = 11. & x(t) = 1.3844 & y(t) = .20276 \\
t = 12. & x(t) = 1.4031 & y(t) = .16875 \\
t = 13. & x(t) = 1.4196 & y(t) = .13899 \\
t = 14. & x(t) = 1.4340 & y(t) = .11341 \\
t = 15. & x(t) = 1.4463 & y(t) = .091778 \\
t = 16. & x(t) = 1.4566 & y(t) = .073750 \\
t = 17. & x(t) = 1.4652 & y(t) = .058910 \\
t = 18. & x(t) = 1.4723 & y(t) = .046823 \\
t = 19. & x(t) = 1.4780 & y(t) = .037063 \\
t = 20. & x(t) = 1.4826 & y(t) = .029240
\end{bmatrix}$$

We see that the solution is converging to the equilibrium point at (1.5, 0).

(e)

There is no peaceful coexistence because almost all the trajectories tend toward an equilibrium point representing a positive population of one species and a zero population of the other. For example, in part (d) we saw that the population of x tends to 1.5 and the population of y tends to 0. The only nonzero initial population distribution which results in no change is x = 1, y = 1.

Looking back to the phase portrait in part (c), the separatrices lie in the corridor separating the curves which bend toward the equilibrium (1.5, 0) and those which bend toward (0, 4). This corridor extends from the origin to the saddle point (1, 1) and from there toward the upper right of the plot. We might guess that the separatrices lie near the line y = x.

```
>  appsep := plot(x, x = 0..3, y = 0..5):
display({portrait, appsep});
```

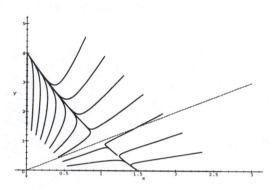

That's not quite right; the separatrix cannot cross another solution curve. We need to draw in something which curves upward as x increases.

```
>  appsep := plot(x^1.3, x = 0..3, y = 0..5):
display({portrait, appsep});
```

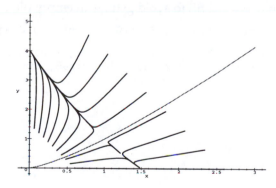

That looks much better. There are two separatrices depicted (approximately) in the above plot, one which approaches (1, 1) from near the origin and one which approaches (1, 1) from the upper right of the plot.

(f)

We want to focus in on the place where the separatrix crosses the line x = 1.5. We can see from the plot above that the y-value of the intersection is between 1 and 2.

```
>   initials := seq([x(0) = 1.5, y(0) = 1 + 0.1*j], j = 0..10):
```

```
>   DEplot([dsys], [x(t), y(t)], t = 0..15, {initials}, x = 0.5..1.5,
y = 0..2, stepsize = 0.5, method = rkf45, arrows = NONE,
linecolor = black);
```

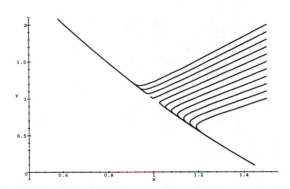

We see that the separatrix lies near the fourth curve from the top, corresponding to y = 1.7. Let's focus in a little more, and use a smaller stepsize. Since our initial conditions are well outside the range we now specify for the graph, we must use the option obsrange = FALSE to avoid getting an empty plot.

```
>   initials := seq([x(0) = 1.5, y(0) = 1.65 + 0.01*j] ,j = 0..10):
```

```
>   DEplot([dsys], [x(t), y(t)], t = 0..20, {initials}, x = 0.9..1.1,
y = 0.9..1.1, stepsize = 0.2, method = rkf45, arrows = NONE,
linecolor = black, obsrange = FALSE);
```

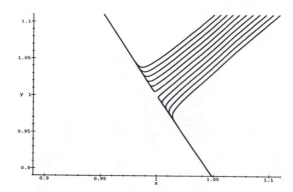

We see that the separatrix is between the fifth and sixth curves, so we conclude that the line $x = 1.5$ cuts the separatrix between $y = 1.69$ and $y = 1.70$. Of course we cannot numerically determine an exact value on the separatrix because our numerical procedure will always introduce errors.

Index

The index uses the same conventions for fonts that are used throughout the book. *Maple* commands, such as **dsolve**, are printed in boldface. Menu options, such as `File`, are printed in a monospaced typewriter font. Trademarked names, such as *Macintosh*, are printed in a slanted font. Everything else is printed in a standard font.